T0181231

Formal Matrices

Algebra and Applications

Volume 23

Algebra and Applications aims to publish well written and carefully refereed monographs with up-to-date information about progress in all fields of algebra, its classical impact on commutative and noncommutative algebraic and differential geometry, K-theory and algebraic topology, as well as applications in related domains, such as number theory, homotopy and (co)homology theory, physics and discrete mathematics.

Particular emphasis will be put on state-of-the-art topics such as rings of differential operators, Lie algebras and super-algebras, group rings and algebras, C^*-algebras, Kac-Moody theory, arithmetic algebraic geometry, Hopf algebras and quantum groups, as well as their applications. In addition, Algebra and Applications will also publish monographs dedicated to computational aspects of these topics as well as algebraic and geometric methods in computer science.

More information about this series at http://www.springer.com/series/6253

Piotr Krylov · Askar Tuganbaev

Formal Matrices

 Springer

Piotr Krylov
Tomsk State University
Tomsk
Russia

Askar Tuganbaev
National Research University MPEI
Moscow
Russia

The study is supported by the Russian Science Foundation (project no. ∼ 16-11-10013)

ISSN 1572-5553　　　　　　　ISSN 2192-2950　(electronic)
Algebra and Applications
ISBN 978-3-319-85272-0　　　ISBN 978-3-319-53907-2　(eBook)
DOI 10.1007/978-3-319-53907-2

Mathematics Subject Classification (2010): 16D20, 16D40, 16D50, 16D90, 16E20, 16E60, 15A15

Printed on acid-free paper

This Springer imprint is published by Springer Nature
The registered company is Springer International Publishing AG
The registered company address is: Gewerbestrasse 11, 6330 Cham, Switzerland

Contents

Symbols

E_{ij}	Matrix unit
$|A|$	Determinant of a matrix A
$d(A)$	Determinant of a formal matrix A
$A \otimes B$	Kronecker product of matrices A and B
$M(n, R)$	Ring of all $n \times n$ matrices over the ring R
$M(n, R, \{s_{ijk}\})$ or $M(n, R, \Sigma)$	Ring of all formal matrices of order n over the ring R with multiplier system $\{s_{ijk}\} = \Sigma$
$M(n, R, s)$	Ring of all formal matrices of order n over the ring R with multiplier s
R°	Opposite ring to a ring R
$P(R)$	Prime radical of the ring R
$J(R)$	Jacobson radical of the ring R
$C(R)$	Center of the ring R
$Q(R)$	Maximal left ring of fractions of the ring R
$R \times S$	Direct product of rings R and S
$A_1 \oplus \ldots \oplus A_n$	Direct sum of modules $A_1, \ldots, A_n$
$\text{Ker } \varphi$ or $\text{Ker}(\varphi)$	Kernel of the homomorphism φ
$\text{Im } \varphi$ or $\text{Im}(\varphi)$	Image of the homomorphism φ
$\text{Rad } A$	Radical of the module A
$\text{Soc } A$	Socle of the module A
$\overline{Z}$	Closure of the submodule Z of the module A
$\hat{A}$	Injective hull of the module A
A^*	Character module of the module A
$\varinjlim_I A_i$	Direct limit of modules A_i
R-Mod (Mod-R)	Category of left (right) modules over the ring R
$P(R)$	Category of finitely generated projective R-modules
$P(A)$	Category of finitely A-projective modules
$\text{End}_R A$ or $\text{End}_R(A)$	Endomorphism ring of an R-module A
$\text{Biend}_R A$	Biendomorphism ring of an R-module A
$\text{End} G$	Endomorphism ring of an Abelian group G

$\mathrm{Hom}_R(A, B)$ — Homomorphism group from an R-module A into an R-module B

$\mathrm{Hom}(G, H)$ — Homomorphism group from an Abelian group G into an Abelian group H

$M \otimes_R N$ — Tensor product of a right R-module M on a left R-module N

$K_0(R)$ — Grothendieck group of the ring R

$K_0(A)$ — Grothendieck group of the category of finitely A-projective modules

$K_1(R)$ — Whitehead group of the ring R

Chapter 1
Introduction

Matrices play an important role in both pure and applied mathematics. Matrices with entries in rings have been intensively studied and used (e.g., see [22, 87]); as well as matrices with entries in semirings (e.g., see [39]), Boolean algebras (e.g., see [66]), semigroups and lattices. This book focuses on a more general object: *formal matrices* or, alternatively, *generalized matrices*. What is a formal matrix? Without going into much detail, one might say that it is a matrix with elements in several different rings or bimodules.

In a well-known paper [89], Morita introduced objects which are now called *Morita contexts* or *pre-equivalence situations*. A Morita context $(R, S, M, N, \varphi, \psi)$ consists of two rings R, S and two bimodules M, N which are linked in a certain way by two bimodule homomorphisms φ and ψ. Initially, Morita contexts were introduced to describe equivalences between the categories of modules. They appeared to be a very convenient tool in the study of properties that transfer from one ring R to another ring S; e.g., see [6, 63, 64].

Morita contexts have been studied in numerous papers and books. There exist analogues of Morita contexts for semirings, Hopf algebras and quasi-Hopf algebras, corings and categories. With the use of the Morita context $(R, S, M, N, \varphi, \psi)$, we can naturally construct the matrix ring $\begin{pmatrix} R & M \\ N & S \end{pmatrix}$ with ordinary matrix operations. This ring is called a *ring of a Morita context*, or a *formal matrix ring* (of order 2), or a *ring of generalized matrices*. The approach taken in this book is to consider a Morita context as a matrix ring. It is straightforward to define a formal matrix ring of any order n.

Formal matrix rings regularly appear in the study of rings and modules, finite-dimensional algebras (e.g., see [10, 11]) and endomorphism rings of Abelian groups (e.g., see [73]). They play an important role in real analysis, e.g., in connection with operator algebras (e.g., see [17, 31]).

An important special case of formal matrix rings is provided by rings of triangular matrices or, more generally, rings of block triangular matrices. Rings of (block)

© Springer International Publishing AG 2017
P. Krylov and A. Tuganbaev, *Formal Matrices*,
Algebra and Applications 23, DOI 10.1007/978-3-319-53907-2_1

triangular matrices naturally appear in the study of finite-dimensional algebras and operator algebras. Rings of formal triangular matrices also serve as a source of examples of rings with non-symmetrical properties.

This book by no means contains all existing results about formal matrix rings. However, the authors believe that the coverage will be sufficient to provide the reader with a general idea of current research frontiers in this area. At the very least, our hope is to provide a first systematic account of the theory of formal matrix rings and modules over them.

This book has five chapters. Chapter 1 is an introduction. Chapter 2 is devoted to formal matrix rings of order 2; sometimes, rings of arbitrary order n are considered. In Chap. 3, we consider modules over formal matrix rings of order 2. This chapter examines in detail injective, flat, projective and hereditary modules over such rings. In Chap. 4, we introduce and study formal matrix rings over a given ring. A lot of attention is paid to individual properties of matrices. The formal matrix ring of order n over a ring R is the closest, in terms of ring properties, to the ordinary ring $M(n, R)$ of all $n \times n$ matrices over R. At the same time it acquires features that are absent in the ring $M(n, R)$. For example, two formal matrix rings of the same order over a given ring are not necessarily isomorphic; see Theorem 4.5.3. Also, the determinant of a formal matrix A over a commutative ring does not necessarily coincides with the determinant of the transposed matrix A^t; see property (7) in Sect. 4.6. In addition, if F is a field, then by the classical Noether–Skolem theorem, every automorphism of the F-algebra $M(n, F)$ is inner. However, there exist rings of formal matrices of order n over F whose group of outer automorphisms contains the symmetric group S_m for some m with $2 \leq m \leq n$; see Item (a) from Sect. 4.1.

In Chap. 5, the Grothendieck group and the Whitehead group of the formal matrix ring $\begin{pmatrix} R & M \\ N & S \end{pmatrix}$ are expressed in terms of the corresponding groups of the rings R and S.

All rings are assumed to be associative and with a non-zero identity element, and modules are assumed to be unitary left modules if not stated otherwise. We write homomorphisms to the left of arguments. The composition of mappings $\alpha \colon X \to Y$ and $\beta \colon Y \to Z$ is denoted by $\alpha\beta$. Thus, $(\alpha\beta)(x) = \beta(\alpha(x))$ for all $x \in X$.

Throughout this book, we use standard concepts and notations of ring theory; e.g., see [79, 80, 98, 108, 112]. If R is a ring, then $J(R)$ is the Jacobson radical of R, $C(R)$ is the center of R, R-Mod is the category of all left R-modules, and $M(n, R)$ is the ring of all $n \times n$ matrices over R. If A is an R-module, then $\mathrm{End}_R A$ or $\mathrm{End}_R(A)$ is the endomorphism ring of A.

Chapter 2
Formal Matrix Rings

In this chapter, we define formal matrix rings of order 2 and formal matrix rings of arbitrary order n. Their main properties are considered and examples of such rings are given.

We indicate the relationship between formal matrix rings, endomorphism rings of modules, and systems of orthogonal idempotents of rings.

For formal matrix rings, the Jacobson radical and the prime radical are described. We find when a formal matrix ring is Artinian, Noetherian, regular, unit-regular, and of stable rank 1.

In the last section, clean and k-good matrix rings are considered.

2.1 Construction of Formal Matrix Rings of Order 2

Let R, S be two rings, M an R-S-bimodule and N an S-R-bimodule. We denote by K the set of all matrices of the form

$$\begin{pmatrix} r & m \\ n & s \end{pmatrix}, \quad \text{where } r \in R,\ s \in S,\ m \in M,\ n \in N.$$

With respect to matrix addition, K is an Abelian group. To turn K into a ring, we have to be able to calculate the "product" $mn \in R$ and the "product" $nm \in S$. This is done as follows.

We assume that there are two bimodule homomorphisms $\varphi \colon M \otimes_S N \to R$ and $\psi \colon N \otimes_R M \to S$; to simplify notation, we also write $\varphi(m \otimes n) = mn$ and $\psi(n \otimes m) = nm$ for all $m \in M$ and $n \in N$. Now we can multiply matrices in K similarly to the case of ordinary matrix rings:

© Springer International Publishing AG 2017
P. Krylov and A. Tuganbaev, *Formal Matrices*,
Algebra and Applications 23, DOI 10.1007/978-3-319-53907-2_2

$$\begin{pmatrix} r & m \\ n & s \end{pmatrix} \begin{pmatrix} r_1 & m_1 \\ n_1 & s_1 \end{pmatrix} = \begin{pmatrix} rr_1 + mn_1 & rm_1 + ms_1 \\ nr_1 + sn_1 & nm_1 + ss_1 \end{pmatrix},$$

$$r, r_1 \in R, \quad s, s_1 \in S, \quad m, m_1 \in M, \quad n, n_1 \in N.$$

In the above definition, rm_1, ms_1, nr_1, sn_1 denote the corresponding module products. We also assume that, for all $m, m' \in M$ and $n, n' \in N$, the two additional associativity relations

$$(mn)m' = m(nm'), \qquad (nm)n' = n(mn') \qquad (*)$$

hold. The set K is a ring with the mentioned operations of addition and multiplication. When checking the ring axioms, we have to consider the main properties of the tensor product and the property that φ and ψ are bimodule homomorphisms. We also have the converse: if K is a ring, then the above associativity relations $(*)$ hold. The ring K is called the *formal matrix ring* (of order 2); it is denoted by $\begin{pmatrix} R & M \\ N & S \end{pmatrix}$. The term *ring of generalized matrices* is also used. Sometimes, we simply refers to a "matrix ring".

If $N = 0$ or $M = 0$, then K is a *ring of formal upper or lower triangular matrices*

$$\begin{pmatrix} R & M \\ 0 & S \end{pmatrix} \quad \text{and} \quad \begin{pmatrix} R & 0 \\ N & S \end{pmatrix}, \quad \text{respectively.}$$

To define a ring of formal upper or lower triangular matrices, it is not necessary to use the homomorphisms φ and ψ.

The images I and J of the homomorphisms φ and ψ are ideals of the rings R and S, respectively. They are called the *trace ideals* of the ring K. We say that K is a *ring with zero trace ideals* or a *trivial* ring provided $\varphi = 0 = \psi$, i.e. $I = 0 = J$. Of course, a ring of formal triangular matrices is a ring with zero trace ideals.

We denote by MN (respectively, NM) the set of all finite sums of elements of the form mn (respectively, nm). The relations

$$I = MN, \quad J = NM, \quad IM = MJ, \quad NI = JN$$

hold. How to formulate the problem of studying formal matrix rings? By studying the ring $\begin{pmatrix} R & M \\ N & S \end{pmatrix}$, we mean figuring out how the properties of this ring depend on the properties of the rings R and S, the properties of the bimodules M and N, and the properties of the homomorphisms φ and ψ.

Sometimes, it is convenient to identify certain matrices with corresponding elements. For example, we can identify the matrix $\begin{pmatrix} r & 0 \\ 0 & 0 \end{pmatrix}$ with the element $r \in R$, and so on. We make similar agreements for matrix sets. For example, the set of matrices

$\begin{pmatrix} X & Y \\ 0 & 0 \end{pmatrix}$ can be written in the form (X, Y) or simply X if $Y = 0$. We use similar rules for matrices with zero upper row.

Let T be a ring. In T, we preserve the previous addition and define a new multiplication $\circ$ by the relation $x \circ y = yx$, $x, y \in T$. As a result, we obtain a new ring T° which is called the ring *opposite* to T. One can directly verify that the opposite ring to the ring $\begin{pmatrix} R & M \\ N & S \end{pmatrix}$ is isomorphic to the formal matrix ring $\begin{pmatrix} R^\circ & N \\ M & S^\circ \end{pmatrix}$ where N is considered as an R°-S°-bimodule and M is considered as an S°-R°-bimodule. We also remark that

$$\begin{pmatrix} R & M \\ N & S \end{pmatrix} \cong \begin{pmatrix} S & N \\ M & R \end{pmatrix}.$$

If V is a right T-module, then the relation $tv = vt$, $t \in T$, $v \in V$, defines a structure of a left T°-module on V, and conversely.

If $M = 0 = N$, then the ring K can be identified with the direct product $R \times S$. Usually, we assume that the product $R \times S$ is a matrix ring.

Let K be a formal matrix ring $\begin{pmatrix} R & M \\ N & S \end{pmatrix}$. Conforming to our agreement about representations of matrices, we can represent the relation

$$K = \begin{pmatrix} eKe & eK(1-e) \\ (1-e)Ke & (1-e)K(1-e) \end{pmatrix}, \tag{2.1}$$

where $e = \begin{pmatrix} 1 & 0 \\ 0 & 0 \end{pmatrix}$. Using this approach, the action of the corresponding homomorphisms φ and ψ coincides with the multiplication in the ring K.

In a certain sense, the converse holds. Namely, let an abstract ring T contain a non-zero idempotent e which is not equal to the identity element. We can form the formal matrix ring

$$K = \begin{pmatrix} eTe & eT(1-e) \\ (1-e)Te & (1-e)T(1-e) \end{pmatrix}.$$

The rings T and K are isomorphic. The correspondence

$$t \to \begin{pmatrix} ete & et(1-e) \\ (1-e)te & (1-e)t(1-e) \end{pmatrix}, \quad t \in T,$$

defines the corresponding isomorphism.

Let K be some formal matrix ring represented in the form (2.1). It is not difficult to specify the structure of ideals and factor rings of the ring K; see also the end of Sect. 2.4 and Propositions 4.2.3 and 4.2.4.

If L is an ideal of the ring K, then one can directly verify that L coincides with the set of matrices

$$\begin{pmatrix} eLe & eL(1-e) \\ (1-e)Le & (1-e)L(1-e) \end{pmatrix}$$

where eLe and $(1-e)L(1-e)$ are ideals of the rings R and S, respectively, and $eL(1-e)$ and $(1-e)Le$ are submodules of M and N, respectively. The subgroups appearing in one of four positions in L coincide with the sets of the corresponding components of elements in L.

We form the matrix group

$$\overline{K} = \begin{pmatrix} eKe/eLe & eK(1-e)/eL(1-e) \\ (1-e)Ke/(1-e)Le & (1-e)K(1-e)/(1-e)L(1-e) \end{pmatrix}.$$

In fact, we have a formal matrix ring $\overline{K}$; we consider this matrix ring in the above general sense. The multiplication of matrices in $\overline{K}$ is induced by the multiplication in K. One can directly verify that the mapping

$$K/L \to \overline{K}, \qquad \begin{pmatrix} r & m \\ n & s \end{pmatrix} + L \to \begin{pmatrix} \overline{r} & \overline{m} \\ \overline{n} & \overline{s} \end{pmatrix},$$

is a ring isomorphism where the dash denotes the corresponding residue class.

A concrete formal matrix ring is defined with the use of two bimodule homomorphisms φ and ψ. In general, the choice of another pair of homomorphisms leads to another ring. We can formulate the problem of classifying formal matrix rings depending on the corresponding pairs of bimodule homomorphisms. The following **isomorphism problem** is related to the above problem:

Let K and K_1 be two formal matrix rings with bimodule homomorphisms φ, ψ and φ_1, ψ_1, respectively. How should the homomorphisms φ, ψ and φ_1, ψ_1 be linked for an isomorphism $K \cong K_1$ to exist?

How many formal matrix rings are there? It follows from the above that the class of formal matrix rings coincides with the class of rings possessing non-trivial idempotents (if we assume that direct products of rings are matrix rings).

The class of formal matrix rings also coincides with the class of endomorphism rings of modules which are decomposable into a direct sum. Indeed, let $G = A \oplus B$ be a right module over some ring T. The endomorphism ring of G is isomorphic to the matrix ring $\begin{pmatrix} \mathrm{End}_T A & \mathrm{Hom}_T(B, A) \\ \mathrm{Hom}_T(A, B) & \mathrm{End}_T B \end{pmatrix}$ with ordinary operations of addition and multiplication of matrices (the product of homomorphisms is taken to be their composite). Conversely, for the ring $K = \begin{pmatrix} R & M \\ N & S \end{pmatrix}$, we have the decomposition $K_K = (R, M) \oplus (N, S)$ into a direct sum of right ideals such that the ring $\mathrm{End}_K(K)$ is isomorphic to the ring $\begin{pmatrix} R & M \\ N & S \end{pmatrix}$.

There are many types of rings which are not necessarily of a matrix origin but are close to formal matrix rings. In particular, various rings of triangular matrices appear. In [35], the authors study so-called *trivial extensions* of rings, which are defined as follows. If R is a ring and M is an R-R-bimodule, then we denote by T the direct sum of Abelian groups R and M, $T = \{(r, m) \mid r \in R, m \in M\}$. The group T is turned into a ring if the multiplication is defined by the relation $(r, m)(r_1, m_1) = (rr_1, rm_1 + mr_1)$. This ring is a trivial extension of the ring R with the use of the bimodule M.

Now we consider the ring of triangular matrices $\begin{pmatrix} R & M \\ 0 & R \end{pmatrix}$ and its subring $\Gamma = \left\{ \begin{pmatrix} r & m \\ 0 & r \end{pmatrix} \middle| r \in R, m \in M \right\}$. The rings T and Γ are isomorphic under the correspondence $(r, m) \to \begin{pmatrix} r & m \\ 0 & r \end{pmatrix}$. Thus, trivial extensions consist of triangular matrices.

Every ring of formal triangular matrices $\begin{pmatrix} R & M \\ 0 & S \end{pmatrix}$ is a trivial extension. Indeed, M can be considered as an $(R \times S)$-$(R \times S)$-bimodule if we assume that $(r, s)m = rm$ and $m(r, s) = ms$. We take the trivial extension $T = \{((r, s), m) \mid r \in R, s \in S, m \in M\}$ of the ring $R \times S$. The correspondence $\begin{pmatrix} r & m \\ 0 & s \end{pmatrix} \to ((r, s), m)$ defines an isomorphism from the ring K onto the ring T. However, there exists a class of rings of triangular matrices containing trivial extensions. Let $f : R \to S$ be a ring homomorphism. In the ring $\begin{pmatrix} R & M \\ 0 & S \end{pmatrix}$, all matrices of the form $\begin{pmatrix} r & m \\ 0 & f(r) \end{pmatrix}$ form a subring.

Here is a more general construction of a ring extension; see [94]. Again, let M be an R-R-bimodule and $\Phi : M \otimes_R M \to R$ an R-R-bimodule homomorphism. We define a multiplication in $R \oplus M$ by the relation

$$(r, m)(r_1, m_1) = (rr_1 + \Phi(m \otimes m_1), rm_1 + mr_1).$$

This multiplication is associative if and only if

$$\Phi(m \otimes m_1)m_2 = m\, \Phi(m_1 \otimes m_2) \tag{2.2}$$

for all $m, m_1, m_2 \in M$. In such a case, $R \oplus M$ is a ring. This ring is denoted by $R \times_\Phi M$ and is called a *semi-trivial extension* of the ring R with the use of M and Φ.

Each formal matrix ring is a semi-trivial extension. Let $\begin{pmatrix} R & M \\ N & S \end{pmatrix}$ be a formal matrix ring with the bimodule homomorphisms φ and ψ. Set $T = R \times S$, $V = M \times N$ and consider V as a natural T-T-bimodule. We denote by Φ the T-T-bimodule homomorphism

$$(\varphi, \psi): V \otimes_T V \to T.$$

It satisfies the corresponding relation (2.2). Consequently, we have a semi-trivial extension $T \times_\Phi V$. The rings $T \times_\Phi V$ and $\begin{pmatrix} R & M \\ N & S \end{pmatrix}$ are isomorphic under the correspondence

$$(r, s) + (m, n) \to \begin{pmatrix} r & m \\ n & s \end{pmatrix}.$$

On the other hand, each semi-trivial extension is embedded in a suitable formal matrix ring. Indeed, let $T \times_\Phi V$ be a semi-trivial extension. The relation corresponding to (2.2) is equivalent to the property that $\begin{pmatrix} T & V \\ V & T \end{pmatrix}$ is a formal matrix ring. The bimodule homomorphisms of this ring coincide with Φ. The mapping

$$T \times_\Phi V \to \begin{pmatrix} T & V \\ V & T \end{pmatrix}, \quad (t, v) \to \begin{pmatrix} t & v \\ v & t \end{pmatrix},$$

is a ring embedding of rings. Thus, we can identify $T \times_\Phi V$ with the matrix ring of the form $\begin{pmatrix} t & v \\ v & t \end{pmatrix}$.

Let T be a commutative ring and R, S two T-algebras. Then the ring $K = \begin{pmatrix} R & M \\ N & S \end{pmatrix}$ is a T-algebra. In this case, we say that K is a *formal matrix algebra*.

2.2   Examples of Formal Matrix Rings of Order 2

Here are some examples of formal matrix rings.

(1). Let S be a ring, M a right S-module, $R = \mathrm{End}_S M$, and $M^* = \mathrm{Hom}_S(M, S)$. Then M is an R-S-bimodule and M^* is an S-R-bimodule, where

$$(s\alpha)m = s\alpha(m), \quad (\alpha r)m = \alpha(r(m)),$$
$$\alpha \in M^*, \quad s \in S, \quad r \in R, \quad m \in M.$$

There exist an R-R-bimodule homomorphism $\varphi: M \otimes_S M^* \to R$ and an S-S-bimodule homomorphism $\psi: M^* \otimes_R M \to S$, which are defined by the formulas

$$\left(\varphi\left(\sum m_i \otimes \alpha_i\right)\right)(m) = \sum m_i \alpha_i(m), \quad \psi\left(\sum \alpha_i \otimes m_i\right) = \sum \alpha_i(m_i),$$

where $m_i, m \in M$ and $\alpha_i \in M^*$. We obtain a matrix ring $\begin{pmatrix} R & M \\ M^* & S \end{pmatrix}$, since the two required associativity relations $(*)$ from Sect. 2.1 hold for φ and ψ.

(2). Let X and Y be a left and a right ideal of the ring R, respectively. Further, let S be any subring in R with $YX \subseteq S \subseteq X \cap Y$. Then $\begin{pmatrix} R & X \\ Y & S \end{pmatrix}$ is a formal matrix ring in which the actions of the mappings φ and ψ are the restrictions of multiplication in R. As a special case, we obtain the ring $\begin{pmatrix} R & Re \\ eR & eRe \end{pmatrix}$, where e is an idempotent.

(3). Let R be a ring, Y a right ideal of R, and S any subring in R containing Y as an ideal. Then S is called a *subidealizer* of the ideal Y in R, and $\begin{pmatrix} R & R \\ Y & S \end{pmatrix}$ is a formal matrix ring.

(4) **Endomorphism rings of Abelian groups.** If G is an Abelian group and $G = A \oplus B$, then the endomorphism ring End G of G is a formal matrix ring; see Sect. 2.1. Abelian groups provide many interesting useful examples of formal matrix rings. First of all, we have rings of triangular matrices. For example, the endomorphism rings of the groups $\mathbb{Q} \oplus \mathbb{Z}$, $\mathbb{Z}(p^n) \oplus \mathbb{Z}$ and $\mathbb{Z}(p^\infty) \oplus \mathbb{Q}$ are isomorphic to the rings $\begin{pmatrix} \mathbb{Q} & \mathbb{Q} \\ 0 & \mathbb{Z} \end{pmatrix}$, $\begin{pmatrix} \mathbb{Z}_{p^n} & \mathbb{Z}_{p^n} \\ 0 & \mathbb{Z} \end{pmatrix}$, and $\begin{pmatrix} \widehat{\mathbb{Z}}_p & A_p \\ 0 & \mathbb{Q} \end{pmatrix}$ respectively, where $\widehat{\mathbb{Z}}_p$ is the ring of p-adic integers and A_p is the field of p-adic numbers.

The endomorphism ring of the p-group $\mathbb{Z}(p^n) \oplus \mathbb{Z}(p^m)$, $n < m$, is an informative illustration of the notion of a formal matrix ring. We can identify this ring with the formal matrix ring $\begin{pmatrix} \mathbb{Z}_{p^n} & \mathbb{Z}_{p^n} \\ \mathbb{Z}_{p^n} & \mathbb{Z}_{p^m} \end{pmatrix}$. We denote this ring by K or $\begin{pmatrix} R & M \\ N & S \end{pmatrix}$.

How can we multiply matrices in the ring K? First of all, we remark that $\mathbb{Z}_{p^n} = \mathbb{Z}_{p^m}/(p^{m-n} \cdot 1)$. Therefore, the rings R and S act on M and N by ordinary multiplication, in a uniquely possible way. Then we pass to the homomorphisms $\varphi \colon M \otimes_S N \to R$ and $\psi \colon N \otimes_R M \to S$. If we consider K as the initial endomorphism ring, then the action of φ and ψ is reduced to the composition of the corresponding homomorphisms. Taking this into consideration, we obtain the following. If $\bar{a} \in M$ and $\bar{b} \in N$, where the dash denotes the residue class, then

$$\varphi(\bar{a} \otimes \bar{b}) = \bar{a} \circ \bar{b} = p^{m-n}\overline{ab}.$$

Then we have

$$\psi(\bar{b} \otimes \bar{a}) = \bar{b} \circ \bar{a} = p^{m-n}\overline{ba},$$

where the last symbols $\bar{b}$ and $\bar{a}$ denote the residue classes in $\mathbb{Z}_{p^m}$ with representatives b and a, respectively.

The trace ideals I and J of the ring K are equal to the ideal $(p^{m-n} \cdot 1)$ of the ring $\mathbb{Z}_{p^n}$ and the ideal $(p^{m-n} \cdot 1)$ of the ring $\mathbb{Z}_{p^m}$, respectively. Therefore, $I \subseteq J(R)$ and $J \subseteq J(S)$. There exists a surjective homomorphism $e \colon S \to R$, $e(\bar{y}) = \bar{y}$, $\bar{y} \in S$. In addition, $\mathrm{Ker}(e) \subseteq J(S)$ and the relation $e(\bar{b} \circ \bar{a}) = \bar{a} \circ \bar{b}$ hold.

In the paper [28], the case $n = 1$ and $m = 2$ is considered in detail. In the ring K, all invertible matrices are described. This is used to construct cryptosystems.

(5) Full matrix rings. Let R be a ring. The full matrix ring $M(n, R)$ can be represented in the form of a formal matrix ring of order 2

$$M(n, R) = \begin{pmatrix} R & M(1 \times (n-1), R) \\ M((n-1) \times 1, R) & M(n-1, R) \end{pmatrix}.$$

This ring provides an example of a ring of block matrices. A more general situation will appear in the proof of Proposition 2.3.3 and in the first paragraph after the proof of Proposition 2.3.3.

(6) see [29]. Let R be a ring, G a finite subgroup of the automorphism group of the ring R and R^G the ring of invariants of the ring R, i.e. R^G is the subring $\{x \in R \mid x^g = x \text{ for all } g \in G\}$. We consider the skew group ring $R * G$ consisting of all formal sums of the form $\sum_{g \in G} r_g g, r_g \in R$. The sums are added component-wise. For multiplication, we use the distributivity law and the relation $rg \cdot sh = rs^{g^{-1}} gh$ for all $r, s \in R$ and $g, h \in G$. It is clear that R is a left R^G-module and a right R^G-module. We consider R as a left and right $R * G$-module as follows:

$$x \cdot r = \sum_{g \in G} r_g r^{g^{-1}}, \quad r \cdot x = \sum_{g \in G} (rr_g)^g$$

for any elements $x = \sum_{g \in G} r_g g \in R * G$ and $r \in R$. The mappings

$$\varphi: R \otimes_{R*G} R \to R^G \quad \text{and} \quad \psi: R \otimes_{R^G} R \to R * G$$

are defined with the use of the relations

$$\varphi(x \otimes y) = \sum_{g \in G} (xy)^g \quad \text{and} \quad \psi(y \otimes x) = \sum_{g \in G} yx^{g^{-1}} g$$

respectively.

The two required associativity conditions $(*)$ from Sect. 2.1 hold, and we eventually obtain the ring $\begin{pmatrix} R^G & R \\ R & R * G \end{pmatrix}$.

2.3 Formal Matrix Rings of Order $n \geq 2$

We make several remarks about formal matrix rings of arbitrary order n. The case $n = 2$, considered in Sect. 2.1, is sufficient to understand how to define such rings.

We fix a positive integer $n \geq 2$. Let $R_1, \ldots, R_n$ be rings, M_{ij} R_i-R_j-bimodules and $M_{ii} = R_i$, $i, j = 1, \ldots, n$. We assume that for any $i, j, k = 1, \ldots, n$ such that

$i \neq j$, $j \neq k$, there is an R_i-R_k-bimodule homomorphism

$$\varphi_{ijk} \colon M_{ij} \otimes_{R_j} M_{jk} \to M_{ik}.$$

For subscripts $i = j$ and $j = k$, we assume that φ_{iik} and φ_{ikk} are canonical isomorphisms

$$R_i \otimes_{R_i} M_{ik} \to M_{ik}, \quad M_{ij} \otimes_{R_j} R_j \to M_{ij}.$$

Instead of $\varphi_{ijk}(a \otimes b)$, we write $a \circ b$ or simply ab. Using this notation, we also assume that $(ab)c = a(bc)$ for all elements $a \in M_{ij}$, $b \in M_{jk}$, $c \in M_{k\ell}$ and subscripts i, j, k, ℓ.

We denote by K the set of all matrices (a_{ij}) of order n with values in the bimodules M_{ij}. The set K is a ring with standard matrix operations of addition and multiplication. We represent K in the form

$$\begin{pmatrix} R_1 & M_{12} & \ldots & M_{1n} \\ M_{21} & R_2 & \ldots & M_{2n} \\ \ldots & \ldots & \ldots & \ldots \\ M_{n1} & M_{n2} & \ldots & R_n \end{pmatrix}. \tag{2.3}$$

We say that K is a *formal matrix ring of order n*. If $M_{ij} = 0$ for all i, j with $i < j$ (resp., $j < i$), then we say that K is a *ring of formal lower* (resp., *upper*) *triangular matrices*.

For every $k = 1, \ldots, n$, set

$$I_k = \sum_{i \neq k} \mathrm{Im}(\varphi_{kik}), \quad \text{where} \quad \varphi_{kik} \colon M_{ki} \otimes_{R_i} M_{ik} \to R_k.$$

Equivalently, $I_k = \sum_{i \neq k} M_{ki} M_{ik}$, where $M_{ki} M_{ik}$ is the set of all finite sums of elements of the form ab, $a \in M_{ki}$, $b \in M_{ik}$. Then I_k is an ideal of the ring R_k. The ideals $I_1, \ldots, I_n$ are called the *trace ideals* of the ring K.

For a better understanding of the structure of formal matrix rings, we determine their interrelations with idempotents and endomorphism rings.

Proposition 2.3.1 *A ring K is a formal matrix ring of order $n \geq 2$ if and only if K contains a complete orthogonal system consisting of n non-zero idempotents.*

Proof If K is a formal matrix ring of order n, then the matrix units $E_{11}, \ldots, E_{nn}$ (see Sect. 4.1) form the required system of idempotents of the ring K.

Conversely, if $\{e_1, \ldots, e_n\}$ is a complete orthogonal system of non-zero idempotents of some ring T, then T is isomorphic to the ring of formal matrices

$$\begin{pmatrix} e_1 T e_1 & e_1 T e_2 & \dots & e_1 T e_n \\ e_2 T e_1 & e_2 T e_2 & \dots & e_2 T e_n \\ \dots & \dots & \dots & \dots \\ e_n T e_1 & e_n T e_2 & \dots & e_n T e_n \end{pmatrix};$$

see Sect. 2.1 in connection to such a ring. $\square$

The case of the direct sums of two modules, examined in Sect. 2.1, can be extended to the direct sums of any finite number of summands.

Proposition 2.3.2 *The class of formal matrix rings of order n coincides with the class of endomorphism rings of modules which are decomposable into a direct sum of n non-zero summands.*

Formal matrix rings of any order n appear in concrete problems. Formal matrix rings of order 2 are usually studied in the general theory; this case is considered mainly because of technical convenience, the case $n > 2$ can be reduced in some sense to the case of matrices of order 2.

Proposition 2.3.3 *A formal matrix ring of order $n > 2$ is isomorphic to some formal matrix ring of order k for every $k = 2, \dots, n - 1$.*

Proof The assertion becomes quite clear if we consider the representation of matrix rings with the use of idempotents or endomorphism rings; see Proposition 2.3.1 and 2.3.2. It is sufficient to "enlarge" in some way idempotents or direct summands. Of course, there is also a direct proof. For example, take $k = 2$. We introduce the following notation for sets of matrices. Set $R = R_1$, $M = (M_{12}, \dots, M_{1n})$,

$$N = \begin{pmatrix} M_{21} \\ \vdots \\ M_{n1} \end{pmatrix}, \quad S = \begin{pmatrix} R_2 & M_{23} & \dots & M_{2n} \\ \dots & \dots & \dots & \dots \\ M_{n2} & M_{n3} & \dots & R_n \end{pmatrix}.$$

Here S is a formal matrix ring of order $n - 1$, M is an R-S-bimodule, N is an S-R-bimodule, and the module multiplications are defined as products of rows and columns on matrices. The homomorphisms φ_{ijk} defining multiplication in K induce the bimodule homomorphisms $\varphi \colon M \otimes_S N \to R$ and $\psi \colon N \otimes_R M \to S$. In addition, the two required associativity relations $(*)$ from Sect. 2.1 hold. As a result, we have the formal matrix ring $\begin{pmatrix} R & M \\ N & S \end{pmatrix}$ and the isomorphism $K \cong \begin{pmatrix} R & M \\ N & S \end{pmatrix}$. The isomorphism is obtained by decomposition of each matrix into four blocks:

$$\begin{pmatrix} a_{11} & a_{12} & \dots & a_{1n} \\ a_{21} & a_{22} & \dots & a_{2n} \\ \dots & \dots & \dots & \dots \\ a_{n1} & a_{n2} & \dots & a_{nn} \end{pmatrix} \to \begin{pmatrix} (a_{11}) & (a_{12} \dots a_{1n}) \\ \begin{pmatrix} a_{21} \\ \dots \\ a_{n1} \end{pmatrix} & \begin{pmatrix} a_{22} \dots a_{2n} \\ \dots \dots \dots \\ a_{n2} \dots a_{nn} \end{pmatrix} \end{pmatrix}. \qquad \square$$

In the proof of the proposition, we actually obtain that formal matrices can be decomposed into blocks, similar to ordinary matrices, i.e., we can represent formal matrices in the form of block matrices. Actions over block matrices are similar to the actions in the case where we have elements instead of blocks. Multiplication of block matrices of the same order is always possible if the factors have the same block decompositions.

Thus, any formal matrix ring can be considered as a ring of (formal) block matrices. Rings of block upper (lower) triangular matrices naturally appear. Rings of (formal) block matrices are used in the theory of finite-dimensional algebras. In particular, rings of block triangular matrices over fields naturally appear in this theory; see [11].

There are a number of constructions which allow us to construct formal matrix rings of larger order starting from formal matrix rings of smaller order. We begin by considering a first method.

Assume that we have a formal matrix ring of the form (2.3). We fix some sequence of positive integers $s_1, \ldots, s_n$ and denote by $\overline{M}_{ij}$ the set of $s_i \times s_j$ matrices with elements in M_{ij} (we recall that $M_{ii} = R$). Further, let $\overline{K}$ be the set of all block matrices $(\overline{M}_{ij})$, $i, j = 1, \ldots, n$. On these matrices, we define operations of addition and multiplication as usual, i.e. the addition is component-wise and $A_{ij} \cdot A_{jk} \in \overline{M}_{ik}$ for any matrices $A_{ij} \in \overline{M}_{ij}$, $A_{jk} \in \overline{M}_{jk}$. Then $\overline{K}$ is turned into a ring of formal block matrices; in addition, $\overline{K}$ is a formal matrix ring of order $s_1 + \ldots + s_n$.

Subsequently, we will use a second simple method of construction of formal matrix rings of larger order. Assume that we have a formal matrix ring of order 2,

$$K = \begin{pmatrix} R & M \\ N & S \end{pmatrix}.$$

We show that there exists a formal matrix ring

$$K_4 = \begin{pmatrix} K & \begin{pmatrix} M \\ S \end{pmatrix} \\ (N\ S) & S \end{pmatrix}.$$

First of all, $\begin{pmatrix} M \\ S \end{pmatrix}$ is a natural K-S-bimodule and $(N\ S)$ is an S-K-bimodule. The mapping

$$\varphi \colon \begin{pmatrix} M \\ S \end{pmatrix} \otimes_S (N\ S) \to K, \qquad \begin{pmatrix} m \\ x \end{pmatrix} \otimes (n, y) \to \begin{pmatrix} mn & my \\ xn & xy \end{pmatrix},$$

is a K-K-bimodule homomorphism and the mapping

$$\psi \colon (N\ S) \otimes_K \begin{pmatrix} M \\ S \end{pmatrix} \to S, \qquad (n, y) \otimes \begin{pmatrix} m \\ x \end{pmatrix} \to nm + yx,$$

is an S-S-bimodule homomorphism. For φ and ψ from Sect. 2.1, the two familiar associativity identities (∗) from Sect. 2.1 hold. Consequently, the specified ring K_4 exists. The same method is used to define the ring

$$K_2 = \begin{pmatrix} K & \begin{pmatrix} R \\ N \end{pmatrix} \\ (R\ M) & R \end{pmatrix}.$$

Now we remark that, in addition to the ring K, there always exists a ring $L = \begin{pmatrix} S & N \\ M & R \end{pmatrix}$. These rings are isomorphic under the correspondence

$$\begin{pmatrix} r & m \\ n & s \end{pmatrix} \rightarrow \begin{pmatrix} s & n \\ m & r \end{pmatrix}.$$

Therefore, in addition to the rings K_2 and K_4, there exist two rings K_1 and K_3 which are isomorphic to K_2 and K_4, respectively. However, we can also construct them directly.

We temporarily turn our attention to the ring of upper triangular matrices of order 3 (they will appear again in Sect. 3.1). Such a ring Γ can be represented in the form $\begin{pmatrix} R & M & L \\ 0 & S & N \\ 0 & 0 & T \end{pmatrix}$, where R, S, T are three rings, M is an R-S-bimodule, L is an R-T-bimodule, and N is an S-T-bimodule. Among the bimodule homomorphisms, only $M \otimes_S N \rightarrow L$ and natural isomorphisms of the form $R \otimes_R M \rightarrow M$ are non-zero. There are two ways to turn Γ into a ring of triangular matrices of order 2. By the first method,

$$\begin{pmatrix} r & m & \ell \\ 0 & s & n \\ 0 & 0 & t \end{pmatrix} \hookrightarrow \begin{pmatrix} \begin{pmatrix} r & m \\ 0 & s \end{pmatrix} & \begin{pmatrix} \ell \\ n \end{pmatrix} \\ (0\ \ 0) & (t) \end{pmatrix}.$$

In this case $\begin{pmatrix} L \\ N \end{pmatrix}$ is a $\begin{pmatrix} R & M \\ 0 & S \end{pmatrix}$-$T$-bimodule. By the second method, (M, L) is an R-$\begin{pmatrix} S & N \\ 0 & T \end{pmatrix}$-bimodule.

2.4 Some Ideals of Formal Matrix Rings

For a formal matrix ring of order n, we find the Jacobson radical and the prime radical. First, we consider the case $n = 2$.

Assume that we have a ring $K = \begin{pmatrix} R & M \\ N & S \end{pmatrix}$. We define four subbimodules of the bimodules M and N. Set

$$J_\ell(M) = \{m \in M \mid Nm \subseteq J(S)\}, \qquad J_r(M) = \{m \in M \mid mN \subseteq J(R)\},$$
$$J_\ell(N) = \{n \in N \mid Mn \subseteq J(R)\}, \qquad J_r(N) = \{n \in N \mid nM \subseteq J(S)\}.$$

Now we form the following sets of matrices:

$$J_\ell(K) = \begin{pmatrix} J(R) & J_\ell(M) \\ J_\ell(N) & J(S) \end{pmatrix}, \qquad J_r(K) = \begin{pmatrix} J(R) & J_r(M) \\ J_r(N) & J(S) \end{pmatrix}.$$

One can directly verify that $J_\ell(K)$ and $J_r(K)$ are a left ideal and a right ideal of the ring K, respectively.

Theorem 2.4.1 ([100]) *We have the relations*

$$J_\ell(K) = J(K) = J_r(K).$$

Proof We have $J(K) = \begin{pmatrix} X & B \\ C & Y \end{pmatrix}$, where X, Y are ideals of the rings R and S, respectively, and B and C are subbimodules in M and N, respectively; see Sect. 2.1. The relations

$$X = eJ(K)e = J(eKe) = J(R)$$

hold, where $e = \begin{pmatrix} 1 & 0 \\ 0 & 0 \end{pmatrix}$. Similarly, we have $Y = J(S)$. Then we have

$$B \subseteq J_\ell(M) \cap J_r(M), \quad C \subseteq J_\ell(N) \cap J_r(N).$$

We have proved that $J(K) \subseteq J_\ell(K) \cap J_r(K)$.

Now we take an arbitrary matrix $\begin{pmatrix} r & m \\ n & s \end{pmatrix}$ in $J_r(K)$ and the identity matrix E. The matrices $E - \begin{pmatrix} r & m \\ 0 & 0 \end{pmatrix}$, $E - \begin{pmatrix} 0 & 0 \\ n & s \end{pmatrix}$ are right invertible in K. Their right inverse matrices are $\begin{pmatrix} x & xm \\ 0 & 1 \end{pmatrix}$ and $\begin{pmatrix} 1 & 0 \\ yn & y \end{pmatrix}$ respectively, where x and y are right inverse to $1 - r$ and $1 - s$ respectively. Consequently, the matrices $\begin{pmatrix} r & m \\ 0 & 0 \end{pmatrix}, \begin{pmatrix} 0 & 0 \\ n & s \end{pmatrix}$ and $\begin{pmatrix} r & m \\ n & s \end{pmatrix}$ are contained in $J(K)$. Therefore, $J_r(K) \subseteq J(K)$. Similarly, we have $J_\ell(K) \subseteq J(K)$. □

We have that $J_\ell(M) = J_r(M)$ and $J_\ell(N) = J_r(N)$. We denote these ideals by $J(M)$ and $J(N)$, respectively. Thus, we have the relation $J(K) = \begin{pmatrix} J(R) & J(M) \\ J(N) & J(S) \end{pmatrix}$.

For an arbitrary ring T, the intersection of all prime ideals of T is called the *prime radical*; it is denoted by $P(T)$.

It is well-known that the prime radical of the ring T coincides with the set of all strongly nilpotent elements of T. We recall that an element $a \in T$ is said to be *strongly nilpotent* if each sequence $a_0, a_1, a_2, \ldots$, such that

$$a_0 = a, \quad a_{n+1} \in a_n T a_n, \quad \forall n \in \mathbb{N},$$

is constantly zero from some term onwards.

We define ideals $P_\ell(M)$, $P_r(M)$, $P_\ell(N)$ and $P_r(N)$ which are similar to the ideals $J_\ell(M)$, $J_r(M)$, $J_\ell(N)$ and $J_r(N)$, respectively. We restrict ourselves to the "left-side" case. Set

$$P_\ell(M) = \{m \in M \mid Nm \subseteq P(S)\}, \quad P_r(M) = \{m \in M \mid mN \subseteq P(R)\}.$$

Then, let $P_\ell(K) = \begin{pmatrix} P(R) & P_\ell(M) \\ P_\ell(N) & P(S) \end{pmatrix}$.

Theorem 2.4.2 ([100]) $P_\ell(K) = P(K) = P_r(K)$.

The proof of Theorem 2.4.2 is a verification of the above relations using the definition of strongly nilpotent elements.   □

We denote by $P(M)$ the equal ideals $P_\ell(M)$ and $P_r(M)$; we also denote the equal ideals $P_\ell(N)$ and $P_r(N)$ by $P(N)$.

Now we pass to a formal matrix ring K of any order n of the form (2.3) from Sect. 2.3. For any two subscripts i and j, we define the subbimodules

$$J_\ell(M_{ij}) = \{x \in M_{ij} \mid M_{ji}x \subseteq J(R_j)\}, \qquad J_r(M_{ij}) = \{x \in M_{ij} \mid xM_{ji} \subseteq J(R_i)\}.$$

For $i = j$, we obtain $J_\ell(R_i) = J_r(R_i) = J(R_i)$.

Theorem 2.4.3 *We have the relation*

$$J(K) = \begin{pmatrix} J(R_1) & J_\ell(M_{12}) & \ldots & J_\ell(M_{1n}) \\ J_\ell(M_{21}) & J(R_2) & \ldots & J_\ell(M_{2n}) \\ \ldots & \ldots & \ldots & \ldots \\ J_\ell(M_{n1}) & J_\ell(M_{n2}) & \ldots & J(R_n) \end{pmatrix}. \qquad (*)$$

A similar relation also holds, where the subscript ℓ is replaced by r.

Proof The case $n = 2$ is considered in Theorem 2.4.1. Let K be a formal matrix ring of order $n \geq 3$. We represent K as a ring of block matrices $\begin{pmatrix} R & M \\ N & R_n \end{pmatrix}$, where R is a formal matrix ring of order $n - 1$ and M, N are the corresponding bimodules; see the proof of Proposition 2.3.3. By Theorem 2.4.1, we have $J(K) = \begin{pmatrix} J(R) & J(M) \\ J(N) & J(R_n) \end{pmatrix}$. By the induction hypothesis, the radical $J(R)$ has the required form. We have to show that the set in the right part of relations $(*)$ coincides with $\begin{pmatrix} J(R) & J(M) \\ J(N) & J(R_n) \end{pmatrix}$. It is sufficient to verify that

$$\begin{pmatrix} J_\ell(M_{1\,n}) \\ \ldots \\ J_\ell(M_{n-1\,n}) \end{pmatrix} = J(M) \quad \text{and} \quad (J_\ell(M_{n1}), \ldots, J_\ell(M_{n\,n-1})) = J(N),$$

where $J(M) = \{m \in M \mid Nm \subseteq J(R_n)\}$, $J(N) = \{n \in N \mid Mn \subseteq J(R)\}$.

The required assertion follows from the definition of the subbimodules $J_\ell(M_{ij})$. We just need the following point. If $x \in J_\ell(M_{nj})$, then $M_{in}x \subseteq J_\ell(M_{ij})$ for all distinct $i, j = 1, \ldots, n$. Indeed, we have

$$M_{ji}M_{in}x \subseteq M_{jn}x \subseteq J(R_j).$$

The proof of the analogue of relations $(*)$ for "the subscript r" is symmetrical to the above proof. □

We have that $J_\ell(M_{ij}) = J_r(M_{ij})$ for distinct i and j. We denote this subbimodule by $J(M_{ij})$.

The prime radical $P(K)$ has a similar structure. Similar to the subbimodules $J_\ell(M_{ij})$ and $J_r(M_{ij})$, we define subbimodules $P_\ell(M_{ij})$ and $P_r(M_{ij})$. The following result holds.

Theorem 2.4.4 *We have the relation*

$$P(K) = \begin{pmatrix} P(R_(1) & P_\ell(M_{12}) & \ldots & P_\ell(M_{1n}) \\ P_\ell(M_{21}) & P(R_2) & \ldots & P_\ell(M_{2n}) \\ \ldots & \ldots & \ldots & \ldots \\ P_\ell(M_{n1}) & P_\ell(M_{n2}) & \ldots & P(R_n) \end{pmatrix}$$

and a similar relation in which the subscript ℓ is replaced by the subscript r.

Now consider the structure of ideals of the ring K. The material of Sect. 2.1 concerning ideals and factor rings can be applied to formal matrix rings of any order n. An ideal L of the ring K is of the form

$$\begin{pmatrix} I_1 & A_{12} & \ldots & A_{1n} \\ A_{21} & I_2 & \ldots & A_{2n} \\ \ldots & \ldots & \ldots & \ldots \\ A_{n1} & A_{n2} & \ldots & I_n \end{pmatrix},$$

where I_i is an ideal of the ring R and A_{ij} is a subbimodule in M_{ij}. It is not difficult to determine certain interrelations between the ideals and subbimodules; in one special case, they are given in Sect. 4.2. The set of matrices

$$\begin{pmatrix} R_1/I_1 & M_{12}/A_{12} & \ldots & M_{1n}/A_{1n} \\ M_{21}/A_{21} & R_2/I_2 & \ldots & M_{2n}/A_{2n} \\ \ldots & \ldots & \ldots & \ldots \\ M_{n1}/A_{n1} & M_{n2}/A_{n2} & \ldots & R_n/I_n \end{pmatrix}$$

naturally forms a formal matrix ring which is canonically isomorphic to the factor ring K/L.

Concluding this section, we calculate the center of a formal matrix ring. We recall that the center of some ring T is denoted by $C(T)$.

Lemma 2.4.5 *The center of the formal matrix ring K consists of all diagonal matrices $\mathrm{diag}(r_1, r_2, \ldots, r_n)$ such that $r_i \in C(R_i)$ and $r_i m = m r_j$ for all $m \in M_{ij}$ and distinct i, j.*

Proof It is clear that diagonal matrices with the mentioned structure are contained in $C(K)$.

Now we assume that a matrix $D = (d_{ij})$ is contained in $C(K)$. It follows from the relations $D E_{kk} = E_{kk} D$ that $d_{ik} = 0 = d_{kj}$ for $i \neq k, k \neq j$. Therefore, $d_{ij} = 0$ for $i \neq j$ and D is a diagonal matrix.

Now we fix subscripts i, j and an element $m \in M_{ij}$. Let A_{ij} be the matrix which has m in position (i, j) and zeros in the remaining positions. It follows from the relations $D A_{ij} = A_{ij} D$ that $d_i m = m d_j$. In particular, for $i = j$, we have $d_i \in C(R_i)$. $\square$

2.5   Ring Properties

We find out when a formal matrix ring of order 2 is Artinian, Noetherian, regular, or unit-regular.

Theorem 2.5.1 ([104]) *A formal matrix ring $K = \begin{pmatrix} R & M \\ N & S \end{pmatrix}$ is left Artinian if and only if R, S are left Artinian rings and ${}_R M, {}_S N$ are Artinian modules. Similar assertions for a right Artinian ring K and a left or right Noetherian ring K also hold.*

Proof $\Rightarrow$. Let $A_1 \supseteq A_2 \supseteq \ldots$ be a descending chain of left ideals of the ring R and $C_1 \supseteq C_2 \supseteq \ldots$ a descending chain of R-submodules of the module M. Then we have the following descending chains in K:

$$\begin{pmatrix} A_1 & 0 \\ NA_1 & 0 \end{pmatrix} \supseteq \begin{pmatrix} A_2 & 0 \\ NA_2 & 0 \end{pmatrix} \cdots,$$

$$\begin{pmatrix} 0 & C_1 \\ 0 & NC_1 \end{pmatrix} \supseteq \begin{pmatrix} 0 & C_2 \\ 0 & NC_2 \end{pmatrix} \cdots.$$

By assumption, there exists a subscript n such that $A_n = A_{n+1} = \ldots$ and $C_n = C_{n+1} = \ldots$. Consequently, $_R R$ and $_R M$ are Artinian modules. Similarly, $_S S$ and $_S N$ are Artinian modules.

$\Leftarrow$. The ring K, considered as a left K-module, is a pair $(R \oplus M, N \oplus S)$, where $R \oplus M$ is a left R-module and $N \oplus S$ is a left S-module; the structure of K is considered in Sect. 3.1. Let $L_1 \supseteq L_2 \supseteq \ldots$ be a descending chain of left ideals of the ring K. Making use of the material of Sect. 3.1, we can write $L_k = (X_k, Y_k)$, where X_k is a submodule of the Artinian module $R \oplus M$ and Y_k is a submodule of the Artinian module $N \oplus S$. In addition, the inclusions $X_1 \supseteq X_2 \supseteq \ldots$ and $Y_1 \supseteq Y_2 \supseteq \ldots$ hold. Consequently, these last two chains stabilize. Therefore, the chain of left ideals $L_1 \supseteq L_2 \supseteq \ldots$ also stabilizes. Consequently, K is a left Artinian ring.

The remaining assertions are similarly proved. $\qquad\square$

We say that a ring T *is of stable rank 1* if for any two elements $a, b \in T$ with $aT + bT = T$, there exists an element $z \in T$ such that $a + bz$ is an invertible element.

For any two elements $a, b \in K$, we set

$$\text{diag}(a, b) = \begin{pmatrix} a & 0 \\ 0 & b \end{pmatrix}, \quad T_{12}(a) = \begin{pmatrix} 1 & a \\ 0 & 1 \end{pmatrix}, \quad T_{21}(b) = \begin{pmatrix} 1 & 0 \\ b & 1 \end{pmatrix}.$$

Theorem 2.5.2 ([24]) *A ring* $K = \begin{pmatrix} R & M \\ N & S \end{pmatrix}$ *is of stable rank 1 if and only if the rings R, S are of stable rank 1.*

Proof $\Rightarrow$. Set $e = \begin{pmatrix} 1 & 0 \\ 0 & 0 \end{pmatrix}$. Then $R \cong eKe$, $S \cong (1 - e)K(1 - e)$, and one can directly verify that R and S are of stable rank 1.

$\Leftarrow$. We take arbitrary matrices

$$A = \begin{pmatrix} a_1 & m_1 \\ n_1 & a_2 \end{pmatrix}, \quad B = \begin{pmatrix} b_1 & m_2 \\ n_2 & b_2 \end{pmatrix}, \quad X = \begin{pmatrix} x_1 & m \\ n & x_2 \end{pmatrix}$$

and assume that $AX + B$ is the identity matrix E of the ring K.

The matrix $G = \begin{pmatrix} A & B \\ -E & X \end{pmatrix}$ of order 2 over K is an invertible matrix with inverse $\begin{pmatrix} X & XA - E \\ E & A \end{pmatrix}$. In particular, the relation $a_1 x_1 + m_1 n + b_1 = 1$ holds in the ring R. Since the ring R is of stable rank 1, there exists an element $r \in R$ such that $a_1 + m_1 n r + b_1 r = u$, where u is an invertible element. Then there exists an element $n' \in N'$ such that

$$
G \cdot \left(\left(\begin{pmatrix} 1 & 0 \\ nr & 1 \end{pmatrix} \quad 0 \\ \begin{pmatrix} r & 0 \\ 0 & 0 \end{pmatrix} \quad E \right) \right) = \left(\left(\begin{pmatrix} u & m_1 \\ n' & a_2 \end{pmatrix} \quad B \\ \begin{pmatrix} * & 0 \\ 0 & -1 \end{pmatrix} \quad X \right) \right);
$$

here and hereafter, $(*)$ denotes elements which are not important for us. We can verify that there exist elements $s_1, s_2 \in S$, $n'' \in N$, two invertible diagonal matrices and one matrix $T_{21}(*)$ in $M(2, K)$ such that

$$
\mathrm{diag}(*, *) \cdot G \cdot \mathrm{diag}(*, *) \cdot T_{21}(*) =
$$

$$
\left(\left(\begin{pmatrix} u & m_1 \\ 0 & s_1 \end{pmatrix} \quad \begin{pmatrix} b_1 & m_2 \\ n'' & s_2 \end{pmatrix} \\ \begin{pmatrix} * & 0 \\ 0 & -1 \end{pmatrix} \quad X \right) \right) = H.
$$

Similarly, there exist elements $s_3, s_4 \in S$ and $m', m'' \in M$ such that $s_1 s_3 + n'' m'' + s_2 s_4 = 1$ in S. Since S is of stable rank 1, there exist elements $s, v \in S$ such that the element v is invertible and $s_1 + n'' m'' s + s_2 s_4 s = v$.

There exists a matrix $T_{21}(*)$ such that

$$
H \cdot T_{21}(*) = \left(\left(\begin{pmatrix} u & * \\ 0 & v \end{pmatrix} \quad \begin{pmatrix} b_1 & m_2 \\ n'' & s_2 \end{pmatrix} \\ * \quad X \right) \right) = V.
$$

Since V and $\begin{pmatrix} u & * \\ 0 & v \end{pmatrix}$ are invertible matrices, and V can be represented in the form

$$
V = \mathrm{diag}(*, *) \cdot T_{21}(*) \cdot T_{12}(*),
$$

where the diagonal matrix is invertible.

As a result, we obtain the relation

$$
\mathrm{diag}(*, *) \cdot G \cdot \mathrm{diag}(*, *) \cdot T_{21}(*) \cdot T_{21}(*) = \mathrm{diag}(*, *) \cdot T_{21}(*) \cdot T_{12}(*)
$$

or

$$
G \cdot \mathrm{diag}(*, *) \cdot T_{21}(*) = \mathrm{diag}(*, *) \cdot T_{21}(*) \cdot T_{12}(*).
$$

Consequently, there exist a matrix Y and invertible matrices U, V, W, P in $M(2, K)$ such that

$$
G \cdot \mathrm{diag}(V, W) \cdot T_{21}(Y) = \mathrm{diag}(U, P) \cdot T_{21}(*) \cdot T_{12}(*).
$$

Then we obtain the relations

$$AV + BWY = U, \quad A + B(WYV^{-1}) = UV^{-1},$$

where UV^{-1} is an invertible matrix; which is what we required. $\quad\square$

Here are several well-known definitions.

Up to Theorem 2.5.6, the symbol R denotes an arbitrary ring.

An element $r \in R$ is said to be *regular* if there exists an element $x \in R$ such that $r = rxr$; a ring R is said to be *regular* if every element of R is regular.

An element $r \in R$ is said to be *unit-regular* if there exists an invertible element $v \in R$ with $r = rvr$; an element r is unit-regular if and only if $r = ue_1 = e_2 u$, where e_1, e_2 are idempotents and u is an invertible element. A ring R is said to be *unit-regular* if every element of R is unit-regular.

If $r = rxr$, then the elements rx and xr are idempotents of the ring R. Therefore, one can directly verify that every principal left (right) ideal of a regular ring R is generated by an idempotent. Consequently, every principal left (right) ideal of a regular ring R is a direct summand of the left (right) R-module R.

Lemma 2.5.3 *If R is a ring and y is an element in R such that $a - aya$ is a regular element, then the element a is also regular.*

Proof There exists an element $z \in R$ such that

$$(a - aya)z(a - aya) = a - aya.$$

Set

$$x = z - zay - yaz + yazay + y.$$

One can directly verify that $axa = a$; therefore, a is a regular element. $\quad\square$

If $x \in R$, then $r(x)$ denotes the *right annihilator* of the element x, i.e., $r(x)$ is the right ideal $\{y \in R \mid xy = 0\}$.

Proposition 2.5.4 ([41]) *For a regular ring R, the following conditions are equivalent.*

(1) R is a unit-regular ring.
(2) $(1 - e)R \cong (1 - f)R$ for any idempotents e, $f \in R$ with $eR \cong fR$.
(3) $r(x) \cong R/xR$ for any element $x \in R$.

Proof $(1) \Rightarrow (2)$. Let e and f be idempotents in R with $eR \cong fR$. We have the direct decompositions

$$R_R = eR \oplus (1 - e)R, \quad R_R = fR \oplus (1 - f)R.$$

Let x be an element in R such that $x(1 - e)R = 0$ and left multiplication by x is an isomorphism $eR \to fR$. There exists an invertible element $u \in R$ with $xux = x$. Since $x(ux - 1) = 0$, we have

$$R \subseteq uxR + (1 - e)R = ufR + (1 - e)R.$$

In addition,

$$xuR = xR = fR \quad \text{and} \quad (xu)^2 = xu.$$

Therefore,

$$ufR \cap (1 - e)R = 0 \quad \text{and} \quad R = ufR \oplus u(1 - e)R.$$

Since u is an invertible element, we also have

$$R = ufR \oplus u(1 - f)R.$$

Finally, we obtain

$$(1 - e)R \cong u(1 - f)R \cong (1 - f)R.$$

$(2) \Rightarrow (3)$. Let $x \in R$. There exists an element $y \in R$ with $xyx = x$. Since xy and yx are idempotents, there exist direct decompositions

$$R_R = yxR \oplus r(x) = xR \oplus (1 - xy)R.$$

Left multiplication of the right ideal yxR by x defines an isomorphism $yxR \to xR$. It follows from (2) that

$$r(x) \cong (1 - xy)R \cong R/xR.$$

$(3) \Rightarrow (1)$. For a given element $x \in R$, there exists an element $y \in R$ with $xyx = x$. As above, we have direct decompositions

$$R_R = yxR \oplus r(x) = xR \oplus (1 - xy)R.$$

It follows from (3) that

$$r(x) \cong R/xR \cong (1 - xy)R.$$

Left multiplication of the right ideal yxR by x defines an isomorphism $yxR \to xR$. As a result, there exists an isomorphism $\varphi \colon R_R \to R_R$. Let u be an invertible element in R such that left multiplication by u is the inverse isomorphism to φ. Then $ux = yx$ and $xux = xyx = x$. $\square$

Proposition 2.5.5 ([41]) *A regular ring R is of stable rank 1 if and only if R is a unit-regular ring.*

Proof ⟹. Let $a \in R$. Then $axa = a$ for some $x \in R$ and $R_R = aR \oplus (1 - ax)R$; see the proof of the implication $(2) \Rightarrow (3)$ of Proposition 2.5.4. Since R is of stable rank 1, $a + (1 - ax)y$ is an invertible element for some element $y \in R$. Therefore, $(a + (1 - ax)y)u = 1$, where u is an invertible element. We obtain

$$a = axa = ax(a + (1 - ax)y)ua = axaua = aua.$$

Consequently, R is a unit-regular ring.

⟸. Assume that we have the relation $aR + bR = R$. All principal right ideals of the regular ring R are direct summands in R_R. We take some direct decomposition $R_R = aR \oplus I$ and consider the restriction of the projection $R_R \to I$ to bR. Then $bR = (aR \cap bR) \oplus J$ for some right ideal J. We have a direct decomposition $R_R = aR \oplus J$. We also have a direct decomposition $R_R = L \oplus K$, where $K = r(a)$; see the proof of the implication $(2) \Rightarrow (3)$ of Proposition 2.5.4. Since $L \cong aR$, it follows from Proposition 2.5.4 that $K \cong J$. Consequently, there exists an element $c \in R$ such that $cL = 0$ and left multiplication by c induces an isomorphism from K onto J. Thus, $cR = J \subseteq bR$ and $c = by$ for some element $y \in R$. We also have that left multiplication by a induces an isomorphism from L onto aR. By considering the relations $aK = 0 = cL$, we obtain that left multiplication by $a + c$ induces an isomorphism $L \oplus K = R_R$ onto $aR \oplus J = R_R$. Consequently, $a + by = a + c$ is an invertible element of the ring R.   □

The paper of Brown and McCoy [21] contains an elementary proof of the property that the matrix ring $M(n, R)$ over a regular ring R is a regular ring. We extend this proof to formal matrix rings of order 2. For formal matrix rings of any order, the assertion (1) of the following Theorem 2.5.6 can be proved with the use of Lemma 1.6 in the book [41]; see also [104].

Theorem 2.5.6 *For a formal matrix ring $K = \begin{pmatrix} R & M \\ N & S \end{pmatrix}$ the following assertions hold.*

(1) K is a regular ring if and only if R, S are regular rings and $x \in xNx$, $y \in yMy$ for all $x \in M$ and $y \in N$.

(2) K is a unit-regular ring if and only if R, S are unit-regular rings and $x \in xNx$, $y \in yMy$ for all $x \in M$ and $y \in N$.

Proof (1). ⟹. One can directly verify that R and S are regular rings and $x \in xNx$, $y \in yMy$ for all $x \in M$ and $y \in N$. For any $x \in M$, there exists a matrix $\begin{pmatrix} a & x' \\ y & b \end{pmatrix} \in K$ such that

$$\begin{pmatrix} 0 & x \\ 0 & 0 \end{pmatrix} \begin{pmatrix} a & x' \\ y & b \end{pmatrix} \begin{pmatrix} 0 & x \\ 0 & 0 \end{pmatrix} = \begin{pmatrix} 0 & x \\ 0 & 0 \end{pmatrix}.$$

Therefore, we have $x = xyx \in xNx$. Similarly, we have that $y \in yMy$ for all $y \in N$.

$\Leftarrow$. If $a \in R$, then a' denotes an arbitrary element with $aa'a = a$. The elements b', x', y' are defined similarly, where $b \in S$, $x \in M$ and $y \in N$.

Set $X = \begin{pmatrix} 0 & 0 \\ x' & 0 \end{pmatrix}$ and $B = A - AXA$. The matrix B has the form $\begin{pmatrix} c & 0 \\ z & d \end{pmatrix}$. Then we construct the matrix $Y = \begin{pmatrix} c' & 0 \\ 0 & d' \end{pmatrix}$ and calculate

$$B - BYB = C = \begin{pmatrix} 0 & 0 \\ n & 0 \end{pmatrix}$$

for some $n \in N$. Thus, we take the matrix $Z = \begin{pmatrix} 0 & n' \\ 0 & 0 \end{pmatrix}$ and see that $C = CZC$. This means that C is a regular element in K. By Lemma 2.5.3, B and A are also regular elements.

(2). The assertion follows from (1), Theorem 2.5.2 and Proposition 2.5.5. $\square$

2.6   Additive Problems

As earlier, the Jacobson radical of the ring R is denoted by $J(R)$.

Let R be a ring. An element $r \in R$ is said to be *clean* if it can be represented in the form $r = u + e$, where e is an idempotent and u is an invertible element. The ring R is said to be *clean* if every element of R is clean. Basic information about clean rings is contained in [93, 109, 110].

Invertible or idempotent elements are clean. If $x \in J(R)$, then $x = (x - 1) + 1$, where $x - 1$ is an invertible element. Consequently, x is a clean element. Therefore, every local ring is a clean ring.

Lemma 2.6.1 ([45]) *Let R be a ring and e an idempotent in R such that eRe and $(1 - e)R(1 - e)$ are clean rings. Then R is a clean ring.*

Proof We identify the ring R with the formal matrix ring

$$\begin{pmatrix} eRe & eR(1 - e) \\ (1 - e)Re & (1 - e)R(1 - e) \end{pmatrix};$$

see Sect. 2.1.

Now we take the matrix $A = \begin{pmatrix} a & x \\ y & b \end{pmatrix} \in R$. There exists a decomposition $a = u + f$, where u is an invertible element of the ring eRe and $f = f^2 \in eRe$. Then

$$b - yu^{-1}x \in (1 - e)R(1 - e).$$

Therefore, the ring $(1 - e)R(1 - e)$ contains an idempotent g and an invertible element v with $b - yu^{-1}x = v + g$. We have the relations

$$A = \begin{pmatrix} u + f & x \\ y & v + g + yu^{-1}x \end{pmatrix} = \begin{pmatrix} f & 0 \\ 0 & g \end{pmatrix} + \begin{pmatrix} u & x \\ y & v + yu^{-1}x \end{pmatrix}.$$

Then $\begin{pmatrix} f & 0 \\ 0 & g \end{pmatrix}$ is an idempotent of the ring R. It remains to verify that $\begin{pmatrix} u & x \\ y & v + yu^{-1}x \end{pmatrix}$ is an invertible matrix. We have the relations

$$\begin{pmatrix} e & 0 \\ -yu^{-1} & 1 - e \end{pmatrix} \begin{pmatrix} u & x \\ y & v + yu^{-1}x \end{pmatrix} \begin{pmatrix} e & -u^{-1}x \\ 0 & 1 - e \end{pmatrix} = \begin{pmatrix} u & 0 \\ 0 & v \end{pmatrix},$$

in which the first matrix and the last two matrices are invertible. $\square$

It is easy to prove the following proposition by induction.

Proposition 2.6.2 *Let R be a ring and $e_1, e_2, \ldots, e_n$ pairwise orthogonal idempotents of the ring R such that $1 = e_1 + e_2 + \ldots + e_n$ and all $e_i R e_i$ are clean rings. Then R is a clean ring.*

The following result can be proved with the use of Propositions 2.6.2 and 2.3.1 concerning the correspondence between formal matrix rings and systems of orthogonal idempotents.

Corollary 2.6.3 *Let K be a formal matrix ring of order n and each of the rings $R_1, R_2, \ldots, R_n$ be clean. Then K is a clean ring.*

There is another way to prove Corollary 2.6.3. First, we apply Lemma 2.6.1 to the ring $\begin{pmatrix} R & M \\ N & S \end{pmatrix}$. A formal matrix ring of order $n > 2$ is a ring of formal block matrices of order 2. Therefore, we can use induction on n.

Proposition 2.6.4 ([93]) *A ring R with Jacobson radical $J(R)$ is clean if and only if $R/J(R)$ is a clean ring and idempotents can be lifted modulo $J(R)$.*

Proof $\Rightarrow$. All homomorphic images of clean rings are clean rings.

Let x be an arbitrary element of the ring R. We have $x = u + e$, where u is an invertible element and e is an idempotent. We have the relations

$$u(x - u^{-1}(1 - e)u) = ue + u^2 - u + eu = x^2 - x,$$
$$x - u^{-1}(1 - e)u = u^{-1}(x^2 - x).$$

We denote by x an element $f \in R$ such that $f^2 - f \in J(R)$. It follows from the relations

$$f - u^{-1}(1 - e)u = u^{-1}(f^2 - f) \in J(R)$$

that idempotents can be lifted modulo $J(R)$.

$\Leftarrow$. Let x be an arbitrary element in R. We denote by $\bar{x}$ a natural image of the element x in the factor ring $R/J(R)$. There exist elements $e, u, v \in R$ such that $\bar{e}^2 = \bar{e}$ and

$$\bar{u}\,\bar{v} = \bar{v}\,\bar{u} = \bar{1}, \quad \bar{x} = \bar{u} + \bar{e}.$$

Since idempotents can be be lifted modulo $J(R)$, we can assume that $e^2 = e$. In addition, $uv = 1 + s$ and $vu = 1 + t$ for some elements $s, t \in J(R)$. Since $1 + s$ and $1 + t$ are invertible elements, uv and vu are invertible elements. Therefore, u and v are invertible elements. Then $x = u + r + e$ for some element $r \in J(R)$, and $u + r$ is an invertible element. $\qquad\square$

Let k be a positive integer and R a ring. An element a of R is said to be *k-good* if a is the sum of k invertible elements of the ring R. A ring is said to be *k-good* if all of its elements are k-good.

Proposition 2.6.5 *If K is a formal matrix ring of order n and all the rings $R_1, R_2, \ldots, R_n$ are k-good, then K is a k-good ring.*

Proof We consider only the case $k = 2$. By the argument given after Corollary 2.6.3, we assume that $K = \begin{pmatrix} R & M \\ N & S \end{pmatrix}$.

We consider an arbitrary matrix $X = \begin{pmatrix} r & m \\ n & s \end{pmatrix}$ in the ring K. We denote by A, B and C the matrices

$$\begin{pmatrix} 0 & m \\ 0 & 0 \end{pmatrix}, \quad \begin{pmatrix} r & 0 \\ 0 & s \end{pmatrix}, \quad \begin{pmatrix} 0 & 0 \\ n & 0 \end{pmatrix},$$

respectively. Since R and S are 2-good rings,

$$B = U + V, \quad \text{where } U = \begin{pmatrix} u_1 & 0 \\ 0 & v_1 \end{pmatrix}, \quad V = \begin{pmatrix} u_2 & 0 \\ 0 & v_2 \end{pmatrix}$$

are invertible matrices. Set

$$A' = A + U = \begin{pmatrix} u_1 & m \\ 0 & v_1 \end{pmatrix}, \quad C' = C + V = \begin{pmatrix} u_2 & 0 \\ n & v_2 \end{pmatrix}.$$

Now we obtain

$$X = A + B + C = A' + C',$$

where A' and C' are invertible matrices. $\qquad\square$

There exists an interesting interrelation between clean and 2-good rings.

An element $s \in R$ is called an *involution* if $s^2 = 1$. When we say that 2 *is invertible in R*, we mean that $2 \cdot 1_R$ is an invertible element. Instead of $r \cdot 2^{-1}$, we write $r/2$.

Proposition 2.6.6 ([23]) *Let R be a ring with $2^{-1} \in R$. The ring R is clean if and only if every element x of R is the sum of an invertible element and an involution (consequently, x is a 2-good element).*

Proof We assume that R is a clean ring and $x \in R$. Then $(x + 1)/2 = u + e$, where $e = e^2$ and u is an invertible element. Then $x = 2u + (2e - 1)$, where $2u$ is an invertible element and $(2e - 1)^2 = 1$.

Conversely, let us assume that every element of the ring R is the sum of an invertible element and an involution. For a given $x \in R$, the element $2x - 1$ can be represented in the form $2x - 1 = u + s$, where u is an invertible element and $s^2 = 1$. Then $x = u/2 + (s + 1)/2$, where $u/2$ is an invertible element and $(s + 1)/2$ is an idempotent. $\square$

We recall that I and J denote the trace ideals of the ring $K = \begin{pmatrix} R & M \\ N & S \end{pmatrix}$ defined in Sect. 2.1; namely, $I = MN$ and $J = NM$. For the ring K with $I \subseteq J(R)$ and $J \subseteq J(S)$, some of the above assertions can be reversed. First, we formulate several general facts.

Lemma 2.6.7 *Let K be a formal matrix ring such that $I \subseteq J(R)$ and $J \subseteq J(S)$.*

(1) $J(K) = \begin{pmatrix} J(R) & M \\ N & J(S) \end{pmatrix}$ *and*

$$K/J(K) \cong R/J(R) \times S/J(S).$$

(2) The matrix $\begin{pmatrix} a & x \\ y & b \end{pmatrix}$ is invertible in K if and only if the elements a and b are invertible in R and S, respectively.

Proof **(1)**. The assertion directly follows from Theorem 2.4.1.

(2). The assertion follows from the property that for any invertible element u of some ring T and every $t \in J(T)$, the element $u + t$ is invertible in T. $\square$

A ring T is said to be *directly finite* if $ba = 1$ for any elements a, b with $ab = 1$.

Theorem 2.6.8 (see [104]) *Let $K = \begin{pmatrix} R & M \\ N & S \end{pmatrix}$ be a formal matrix ring such that $I \subseteq J(R)$ and $J \subseteq J(S)$.*

(1) K is a clean ring if and only if R and S are clean rings.
(2) K is a 2-good ring if and only if R and S are 2-good rings.
(3) K is a directly finite ring if and only if R and S are directly finite rings.

Proof It follows from Corollary 2.6.3 and Proposition 2.6.5 that it is sufficient to prove the necessity of the conditions in (1) and (2).

(**1**). $\Rightarrow$. It follows from Proposition 2.6.4 and the isomorphism from Lemma 2.6.7 that $R/J(R)$ is a clean ring. Since K is a clean ring, it follows from the proof of Proposition 2.6.4 that for each matrix X in K, there exists an idempotent matrix W with $X - W \in K(X^2 - X)$. We assume that a similar property holds for all elements of the ring R. Then it follows from the proof of Proposition 2.6.4 that idempotents can be lifted modulo $J(R)$ and R is a clean ring by Proposition 2.6.4. Similarly, it is verified that S is a clean ring.

Thus, let $x \in R$. There exists an idempotent matrix $\begin{pmatrix} a & b \\ c & d \end{pmatrix}$ such that

$$\begin{pmatrix} x & 0 \\ 0 & 0 \end{pmatrix} - \begin{pmatrix} a & b \\ c & d \end{pmatrix} \in K\left(\begin{pmatrix} x^2 & 0 \\ 0 & 0 \end{pmatrix} - \begin{pmatrix} x & 0 \\ 0 & 0 \end{pmatrix}\right) = K\begin{pmatrix} x^2 - x & 0 \\ 0 & 0 \end{pmatrix}.$$

Then

$$\begin{pmatrix} x - a & -b \\ -c & -d \end{pmatrix} \in \begin{pmatrix} R(x^2 - x) & 0 \\ N(x^2 - x) & 0 \end{pmatrix}.$$

Therefore, $b = 0 = d$, whence $a^2 = a$, i.e., a is an idempotent. Then $x - a \in R(x^2 - x)$; which is what we required.

(**2**). $\Rightarrow$. For an element $r \in R$, we have

$$\begin{pmatrix} r & 0 \\ 0 & 0 \end{pmatrix} = \begin{pmatrix} u_1 & a \\ b & v_1 \end{pmatrix} + \begin{pmatrix} u_2 & c \\ d & v_2 \end{pmatrix},$$

where the last two matrices are invertible. Consequently, $r = u_1 + u_2$, where elements u_1 and u_2 are invertible by Lemma 2.6.7.

(**3**). The necessity of the conditions may be directly verified; the inclusions $I \subseteq J(R)$ and $J \subseteq J(S)$ are not required.

Now let $\begin{pmatrix} a & x \\ y & b \end{pmatrix}\begin{pmatrix} a' & x' \\ y' & b' \end{pmatrix} = 1$ in K. Then $aa' + xy' = 1$ in R and $yx' + bb' = 1$ in S. Since $xy' \in J(R)$ and $yx' \in J(S)$, the products aa' and bb' are invertible in R and S, respectively. Since R and S are directly finite rings, a and b are invertible elements. By Lemma 2.6.7, $\begin{pmatrix} a & x \\ y & b \end{pmatrix}$ is an invertible matrix, whence $\begin{pmatrix} a' & x' \\ y' & b' \end{pmatrix}\begin{pmatrix} a & x \\ y & b \end{pmatrix} = 1$. $\qquad\qquad\square$

In Sect. 2.3, the trace ideals $I_1, \ldots, I_n$ of the formal matrix ring K of order n were defined. For such a ring K, we can prove the analogue of Theorem 2.6.8 provided conditions $I_k \subseteq J(R_k)$, $k = 1, \ldots, n$, hold. We can use induction on n by representing K as rings of formal block matrices of order 2.

We present some of Henriksen's interesting results about good ordinary matrix rings $M(n, R)$ over an arbitrary ring R.

Lemma 2.6.9 *For $n > 1$, every diagonal matrix in $M(n, R)$ is 2-good.*

Proof Let $D = \mathrm{diag}(a_1, a_2, \ldots, a_n)$ be a diagonal matrix. We consider the matrices

$$
U = \begin{pmatrix}
a_1 & 0 & 0 & \ldots & 0 & 0 & 1 \\
1 & a_2 & 0 & \ldots & 0 & 0 & 0 \\
0 & 1 & a_3 & \ldots & 0 & 0 & 0 \\
\cdot & \cdot & \cdot & \ldots & \cdot & & \cdot \\
0 & 0 & 0 & \ldots & 1 & a_{n-1} & 0 \\
0 & 0 & 0 & \ldots & 0 & 1 & 0
\end{pmatrix},
$$

$$
V = \begin{pmatrix}
0 & 0 & 0 & \ldots & 0 & 0 & -1 \\
-1 & 0 & 0 & \ldots & 0 & 0 & 0 \\
0 & -1 & 0 & \ldots & 0 & 0 & 0 \\
\cdot & \cdot & \cdot & \ldots & \cdot & \cdot & \cdot \\
0 & 0 & 0 & \ldots & -1 & 0 & 0 \\
0 & 0 & 0 & \ldots & 0 & -1 & a_n
\end{pmatrix}.
$$

With the use of elementary transformations of rows or columns, the matrices U and V can be transformed to the identity matrix. Consequently, U and V are invertible matrices. In addition, $D = U + V$. □

Lemma 2.6.10 (Kaplansky) *For any $n \geq 1$, every matrix in $M(n, R)$ is the sum of a diagonal matrix and an invertible matrix.*

Proof We use induction on n. If $n = 1$ and $a \in R$, then $a = (a - 1) + 1$ and the assertion holds.

We assume that the assertion holds for some $n \geq 1$. If $A' \in M(n + 1, R)$, then the matrix A' can be represented in the following block form:

$$
A' = \begin{pmatrix} A & B \\ C & d \end{pmatrix},
$$

where $A \in M(n, R)$, $d \in R$, B is a column vector and C is a row vector. By the induction hypothesis, $A = D + V$, where $D, V \in M(n, R)$, D is a diagonal matrix and V is an invertible matrix. We consider the matrices

$$
D' = \begin{pmatrix} D & 0 \\ 0 & d - 1 - CV^{-1}B \end{pmatrix}, \quad V' = \begin{pmatrix} V & B \\ C & 1 + CV^{-1}B \end{pmatrix}.
$$

Then $A' = D' + V'$, where D' is a diagonal matrix. Let E be the identity matrix in $M(n, R)$ and

$$
P = \begin{pmatrix} E & 0 \\ -CV^{-1} & 1 \end{pmatrix}, \quad Q = \begin{pmatrix} V^{-1} & V^{-1}B \\ 0 & 1 \end{pmatrix}.
$$

Then $PV'Q$ is the identity matrix in $M(n+1, R)$ and P, Q are invertible matrices in $M(n+1, R)$. Therefore, V' is an invertible matrix; which is what we required.   □

By combining the last two lemmas, we obtain the following result.

Theorem 2.6.11 ([56]) *For $n > 1$, the matrix ring $M(n, R)$ is a 3-good ring for any ring R.*

We can show that $M(n, R)$ is also a 4-good ring.

A matrix ring $M(n, R)$ is not necessarily a 2-good ring. Let $n > 1$, F an arbitrary field and $R = F[x_1, \ldots, x_n]$ the polynomial ring. In [56], it is shown that the ring $M(n, R)$ contains a matrix which is not 2-good.

Of course, there exists a ring R such that $M(n, R)$ is a 2-good ring for any $n > 1$. A ring R is called a *diagonalizable* ring if for any n, each matrix $A \in M(n, R)$ is equivalent to some diagonal matrix D, i.e., $UAV = D$ for some invertible matrices U and V. For example, commutative principal ideal rings and commutative valuation rings are diagonalizable rings.

The following theorem follows from Lemma 2.6.9 and the definition of a diagonalizable ring.

Theorem 2.6.12 *If R is a diagonalizable ring and $n > 1$, then $M(n, R)$ is a 2-good ring.*

Every formal matrix M is the sum of a diagonal matrix and an invertible matrix, i.e., Kaplansky's lemma 2.6.10 holds for M. The proof of Lemma 2.6.10 can be directly extended to formal matrices. For arbitrary formal matrices, the analogue of Lemma 2.6.9 is not true. This follows, for example, from Theorem 2.6.8(2).

In the literature, there are various notions close to those of a clean element and a clean ring. An element $r \in R$ is said to be *strongly clean* if r can be represented in the form $r = u + e$, where u is an invertible element, e is an idempotent, and $ue = eu$. A ring R is said to be *strongly clean* if every element of R is strongly clean.

Strongly clean matrix rings and individual strongly clean matrices over local rings have been studied in numerous papers. The paper [19] contains a complete characterization of commutative local rings R such that $M(n, R)$ is a strongly clean ring. Let R be an arbitrary local ring. In [105, 115], the authors determined when $M(2, R)$ is a strongly clean ring and $M(2, R, s)$ is a strongly clean ring, respectively; see the end of Sect. 4.1 for more about the ring $M(2, R, s)$.

Chapter 3
Modules over Formal Matrix Rings

We set out the basic theory of modules over formal matrix rings and consider the structure of some submodules (small and essential submodules, the socle and the radical). We pay much attention to injective, flat and projective modules over formal matrix rings of order 2.

As an application, we calculate the maximal ring of fractions of the formal matrix ring of order 2 and give examples of Abelian groups whose endomorphism rings are hereditary.

We also establish equivalences between the category of modules over the formal matrix ring of order 2 and categories of modules over original rings.

3.1 Initial Properties of Modules over Formal Matrix Rings

What is the structure of modules over the formal matrix ring $K = \begin{pmatrix} R & M \\ N & S \end{pmatrix}$? We can construct them on the basis of R-modules and S-modules. Let X be an R-module and Y an S-module. We assume that there is an R-module homomorphism $f \colon M \otimes_S Y \to X$ and an S-module homomorphism $g \colon N \otimes_R X \to Y$ such that

$$m(nx) = (mn)x, \quad n(my) = (nm)y, \quad m \in M, \ n \in N, \ x \in X, \ y \in Y,$$

where we assume that nx denotes $g(n \otimes x)$, and my denotes $f(m \otimes y)$. The group of column vectors $\begin{pmatrix} X \\ Y \end{pmatrix}$ is turned into a K-module if we take the product of the matrix by a column as the module multiplication,

© Springer International Publishing AG 2017
P. Krylov and A. Tuganbaev, *Formal Matrices*,
Algebra and Applications 23, DOI 10.1007/978-3-319-53907-2_3

$$\begin{pmatrix} r & m \\ n & s \end{pmatrix} \begin{pmatrix} x \\ y \end{pmatrix} = \begin{pmatrix} rx + my \\ nx + sy \end{pmatrix}.$$

The homomorphisms f and g are called the *homomorphisms of module multiplication*.

Every K-module has the form of a module of columns. More precisely, we can obtain it by the above method. Let V be a K-module and $e = \begin{pmatrix} 1 & 0 \\ 0 & 0 \end{pmatrix}$. Then eV is an R-module, $(1-e)V$ is an S-module, and $\begin{pmatrix} eV \\ (1-e)V \end{pmatrix}$ is a K-module. This becomes clear if we remember the agreements made in Sect. 2.1 concerning the representation of matrices and matrix sets. For example, we can assume that K has the form

$$\begin{pmatrix} eKe & eK(1-e) \\ (1-e)Ke & (1-e)K(1-e) \end{pmatrix}.$$

Then the module multiplication in the mentioned modules naturally holds and the homomorphisms of module multiplication

$$M \otimes_S (1-e)V \to eV, \qquad N \otimes_R eV \to (1-e)V$$

are "restrictions" of the canonical isomorphism $K \otimes_K V \to V$ to the corresponding submodules. The correspondence of elements

$$v \to \begin{pmatrix} ev \\ (1-e)v \end{pmatrix}, \qquad v \in V,$$

is an isomorphism of the K-modules V and $\begin{pmatrix} eV \\ (1-e)V \end{pmatrix}$. In particular, as a left K-module, K has the form $\begin{pmatrix} (R, M) \\ (N, S) \end{pmatrix}$ with homomorphisms of module multiplication

$$m \otimes (n, s) \to (mn, ms), \qquad n \otimes (r, m) \to (nr, nm).$$

For right K-modules, a similar argument holds. Every right K-module has the form of a module of row vectors (X, Y), where X is a right R-module and Y is a right S-module. With this K-module, we associate the module homomorphisms $Y \otimes_S N \to X$ and $X \otimes_R M \to Y$ which satisfy the corresponding associativity relations. The module multiplication is the multiplication of rows by the matrix. All properties of left K-modules have right-side analogues. These properties can also be automatically proved by considering right modules over the opposite ring $K°$ (see Sect. 2.1). It is noted in Sect. 2.1 that the ring $K°$ is also a formal matrix ring.

We pay special attention to the fact that left K-modules of the form $\begin{pmatrix} X \\ Y \end{pmatrix}$ and their elements may also be represented as rows.

For (right or left) K-modules, we agree to write them as matrices which is also valid for the ring K itself. For example, we write X instead of $\begin{pmatrix} X \\ 0 \end{pmatrix}$ and $(X, 0)$; we also write x instead of $\begin{pmatrix} x \\ 0 \end{pmatrix}$ and $(x, 0)$, and so on.

Let (X, Y) be some (left) K-module. Set $MY = \operatorname{Im} f$ and $NX = \operatorname{Im} g$. It is clear that MY and NX are the sets of all finite sums of elements of the form my and nx, respectively. We have the inclusions $IX \subseteq MY$ and $JY \subseteq NX$. If subgroups of M, N, Y, X are considered, we also use notations which are similar to MN, MY, NX.

Instead of the homomorphisms of module multiplication $f\colon M \otimes_S Y \to X$, $g\colon N \otimes_R X \to Y$, it is sometimes more convenient to use the S-homomorphism f' and R-homomorphism g':

$$f'\colon Y \to \operatorname{Hom}_R(M, X), \quad f'(y)(m) = f(m \otimes y) = my, \ y \in Y, \ m \in M,$$
$$g'\colon X \to \operatorname{Hom}_S(N, Y), \quad g'(x)(n) = g(n \otimes x) = nx, \ x \in X, \ n \in N.$$

The homomorphisms f' and g' correspond to the homomorphisms f and g for natural isomorphisms of Abelian groups

$$\operatorname{Hom}_R(M \otimes_S Y, X) \cong \operatorname{Hom}_S(Y, \operatorname{Hom}_R(M, X)),$$
$$\operatorname{Hom}_S(N \otimes_R X, Y) \cong \operatorname{Hom}_R(X, \operatorname{Hom}_S(N, Y)),$$

respectively. In particular, when we define K-modules, we can start from the homomorphisms f' and g'; this is equivalent to the initial approach. Therefore, f' and g' may also be called *homomorphisms of module multiplication*.

The K-module homomorphisms can be represented by pairs consisting of the homomorphisms of R-modules and S-modules. Let us have two K-modules (X, Y) and (X_1, Y_1). We assume that $\alpha\colon X \to X_1$ and $\beta\colon Y \to Y_1$ is an R-homomorphism and an S-homomorphism, respectively, and $\alpha(my) = m\beta(y)$, $\beta(nx) = n\alpha(x)$ for all $m \in M$, $n \in N$, $x \in X$, $y \in Y$. Then the mapping $(X, Y) \to (X_1, Y_1)$, $(x, y) \to (\alpha(x), \beta(y))$ is a K-homomorphism. Now assume that we have some K-homomorphism $\Phi\colon (X, Y) \to (X_1, Y_1)$. It follows from the relations

$$\Phi\left(\begin{pmatrix} 1 & 0 \\ 0 & 0 \end{pmatrix}(x, y)\right) = \begin{pmatrix} 1 & 0 \\ 0 & 0 \end{pmatrix}\Phi(x, y),$$
$$\Phi\left(\begin{pmatrix} 0 & 0 \\ 0 & 1 \end{pmatrix}(x, y)\right) = \begin{pmatrix} 0 & 0 \\ 0 & 1 \end{pmatrix}\Phi(x, y)$$

that Φ acts as $\Phi(x, y) = (\alpha(x), \beta(y))$, where α and β are mappings $X \to X_1$ and $Y \to Y_1$, respectively. We can directly verify that α is an R-homomorphism, β is an S-homomorphism, and $\alpha(my) = m\beta(y)$, $\beta(nx) = n\alpha(x)$ for all elements of these relations. Therefore, we can represent the K-module homomorphisms in the form of pairs (α, β).

The above material on the structure of K-modules can be described in terms of categories. We will prove that the category of K-modules is equivalent to a category of "tetrads". We define the category $\mathcal{A}(K)$. Its objects are expressions (X, Y, f, g), where X is an R-module, Y is an S-module, $f: M \otimes_S Y \to X$ is an R-homomorphism, $g: N \otimes_R X \to Y$ is an S-homomorphism, and the following diagrams are commutative:

$$
\begin{array}{ccc}
M \otimes_S N \otimes_R X \xrightarrow{1 \otimes g} M \otimes_S Y \xrightarrow{f} & X \\
\downarrow \varphi \otimes 1 & \downarrow 1_X \\
R \otimes_R X \xrightarrow{\mu} & X \\
N \otimes_R M \otimes_S Y \xrightarrow{1 \otimes f} N \otimes_R X \xrightarrow{g} & Y \\
\downarrow \psi \otimes 1 & \downarrow 1_Y \\
S \otimes_S Y \xrightarrow{\nu} & Y
\end{array}
\qquad (3.1.1)
$$

where μ and ν are canonical homomorphisms. Morphisms $(X, Y, f, g) \to (X_1, Y_1, f_1, g_1)$ coincide with pairs (α, β), where $\alpha \in \mathrm{Hom}_R(X, X_1), \beta \in \mathrm{Hom}_S(Y, Y_1)$ such that the following diagrams are commutative:

$$
\begin{array}{ccccc}
M \otimes_S Y \xrightarrow{f} X & & N \otimes_R X \xrightarrow{g} Y \\
\downarrow 1 \otimes \beta \quad \downarrow \alpha, & & \downarrow 1 \otimes \alpha \quad \downarrow \beta. & & (3.1.2) \\
M \otimes_S Y_1 \xrightarrow{f_1} X_1 & & N \otimes_R X_1 \xrightarrow{g_1} Y_1
\end{array}
$$

Theorem 3.1.1 ([42, 94]) *The categories K-Mod and $\mathcal{A}(K)$ are equivalent.*

Proof We define the functor $F: K\text{-Mod} \to \mathcal{A}(K)$. For a K-module V, we take $(eV, (1-e)V, f, g)$ as $F(V)$, where $e = \begin{pmatrix} 1 & 0 \\ 0 & 0 \end{pmatrix}$, eV and $(1-e)V$ are the corresponding modules from the beginning of this section, and $f: M \otimes_S (1-e)V \to eV$, $g: N \otimes_R eV \to (1-e)V$ are "restrictions" of the canonical isomorphism $K \otimes_K V \to V$. Diagrams (3.1.1) are commutative. Let $\Phi: V \to W$ be a K-module homomorphism. Then $F(\Phi): F(V) \to F(W)$ is the pair (α, β), where α is the restriction of Φ to eV, and β is the restriction of Φ to $(1-e)V$. We remark that $\Phi(eV) = e\Phi(eV) \subseteq eW$; a similar situation holds for $(1-e)V$. One can directly verify that diagrams (3.1.2) are commutative.

Now we define a functor $G: \mathcal{A}(K) \to K\text{-mod}$. Let $(X, Y, f, g) \in \mathcal{A}(K)$. We assume that $G(X, Y, f, g)$ is the group of row vectors $\{(x, y) \mid X \in X, y \in Y\}$. We define the action of K on $G(X, Y, f, g)$ as

$$\begin{pmatrix} r & m \\ n & s \end{pmatrix} (x, y) = (rx + f(m \otimes y), g(n \otimes x) + sy).$$

As a result, we obtain a K-module $G(X, Y, f, g)$. If $(\alpha, \beta) \colon (X, Y, f, g) \to (X_1, Y_1, f_1, g_1)$ is a morphism in $\mathcal{A}(K)$, then we define $G(\alpha, \beta)$ by the relation

$$G(\alpha, \beta) = (\alpha(x), \beta(y)), \quad x \in X, \ y \in Y.$$

One can directly verify that $G(\alpha, \beta)$ is a K-homomorphism.

It remains to show that the functors F and G determine the equivalence of categories K-Mod and $\mathcal{A}(K)$. Namely, the composition GF is naturally equivalent to the identity functor of the category K-Mod, and the composition FG is naturally equivalent to the identity functor of the category $\mathcal{A}(K)$. We define natural transformations σ and τ as follows. If V is some K-module, then, by our definitions, $GF(V)$ is the K-module $(eV, (1 - e)V)$. The mapping $\sigma_V \colon GF(V) \to V, \sigma_V(ev, (1 - e)v) = v$, $v \in V$, is a K-isomorphism. In addition, if $\varphi \colon V \to W$ is a K-module homomorphism, then $GF(\varphi)\sigma_W = \sigma_V\varphi$. Consequently, σ is a natural equivalence.

Now we take an object $A = (X, Y, f, g)$ of the category $\mathcal{A}(K)$. In fact, $FG(A)$ coincides with A and we can take the identity morphism of the object A as the morphism $\tau_A \colon FG(A) \to A$. If A' is another object of the category $\mathcal{A}(K)$ and $\psi \colon A \to A'$ is a morphism, then one can directly verify that the object diagram

$$
\begin{array}{ccc}
FG(A) & \overset{FG(\psi)}{\longrightarrow} & FG(A') \\
\tau_A \downarrow & & \downarrow \tau_{A'} \\
A & \overset{\psi}{\longrightarrow} & A'
\end{array}
$$

in $\mathcal{A}(K)$ is commutative. Consequently, τ is a natural equivalence and the theorem is proved. $\qquad\square$

There are simple but quite useful constructions of K-modules based on tensor products and Hom groups. Let X be an R-module. The group of row vectors $(X, N \otimes_R X)$ is a K-module (we consider $N \otimes_R X$ as a canonical S-module) in which the homomorphism

$$M \otimes_S (N \otimes_R X) \to X, \qquad m(n \otimes x) \to (mn)x,$$

and the identity automorphism $N \otimes_R X \to N \otimes_R X$ are homomorphisms of module multiplication. Thus, using our notation, we have $m(n \otimes x) = (mn)x$ and $nx = n \otimes x$. Similarly, we can start with an S-module Y and define the K-module $(M \otimes_S Y, Y)$. We will use the notation $T(X) = N \otimes_R X$ and $T(Y) = M \otimes_S Y$. The modules $(X, T(X))$ and $(T(Y), Y)$ satisfy the following specific property.

Lemma 3.1.2 *Assume that we have an R-module X, a K-module (A, B) and an R-homomorphism $\alpha\colon X \to A$. Then there exists a unique S-homomorphism $\beta\colon T(X) \to B$ such that $(\alpha, \beta)\colon (X, T(X)) \to (A, B)$ is a K-module homomorphism.*

A similar assertion holds for an S-module Y, an S-homomorphism $Y \to B$ and K-modules (A, B), (T(Y), Y).

Proof The mapping $N \times X \to B$, $(n, x) \to n\alpha(x)$, $n \in N$, $x \in X$, is S-balanced. Consequently, there exists an S-homomorphism $\beta\colon T(X) \to B$ which acts on generator elements as $\beta(n \otimes x) = n\alpha(x)$. The pair (α, β) determines a K-homomorphism, since $\alpha(m(n \otimes x)) = m\beta(n \otimes x)$ and $\beta(nx) = n\alpha(x)$ for all $m \in M, n \in N, x \in X$.

The uniqueness of β is understood in the sense indicated below. If $(\alpha, \gamma)\colon (X, T(X)) \to (A, B)$ is some K-homomorphism, then $\gamma = \beta$. Indeed, it follows from the definition of the module $(X, T(X))$ that

$$\gamma(n \otimes x) = \gamma(nx) = n\alpha(x) = \beta(n \otimes x), \quad \gamma = \beta.$$

We can use a similar proof for the modules (A, B) and $(T(Y), Y)$. □

Now we form the group of row vectors $(X, \mathrm{Hom}_R(M, X))$, where we consider the group $\mathrm{Hom}_R(M, X)$ as an S-module in a standard way. In fact, we have a K-module with homomorphisms of module multiplication

$$M \otimes_S \mathrm{Hom}_R(M, X) \to X, \quad m \otimes \alpha \to \alpha(m),$$
$$N \otimes_R X \to \mathrm{Hom}_R(M, X), \quad n \otimes x \to \beta, \qquad \text{where}$$
$$\beta(m) = (mn)x, \quad m \in M, \quad n \in N, \quad x \in X, \quad \alpha \in \mathrm{Hom}_R(M, X).$$

In accordance with our notational convention, we have the relations $m\alpha = \alpha(m)$ and $(nx)(m) = (mn)x$. Similarly, the S-module Y provides the K-module $(\mathrm{Hom}_S(N, Y), Y)$ with module multiplications

$$n\gamma = \gamma(n), \quad (my)(n) = (nm)y, \quad n \in N, \quad m \in M, \quad y \in Y, \quad \gamma \in \mathrm{Hom}_S(N, Y).$$

We will write $H(X)$ instead of $\mathrm{Hom}_R(M, X)$ and $H(Y)$ instead of $\mathrm{Hom}_S(N, Y)$. The modules $(X, H(X))$ and $(H(Y), Y)$ have the following important property related to homomorphisms.

Lemma 3.1.3 *Let X be an R-module, (A, B) a K-module and $\alpha\colon A \to X$ an R-homomorphism. We define a mapping $\beta\colon B \to H(X)$ by the relation $\beta(b)(m) = \alpha(mb)$, $b \in B$, $m \in M$. Then β is an S-homomorphism and (α, β) is a K-homomorphism $(A, B) \to (X, H(X))$. Such a homomorphism β is unique.*

A similar assertion holds for the S-module Y, the S-homomorphism $B \to Y$ and the K-modules (A, B), (H(Y), Y).

Proof The mapping β is a homomorphism of Abelian groups. In addition,

$$\beta(sb)(m) = \alpha(m(sb)), \quad (s\beta(b))(m) = \beta(b)(ms) = \alpha((ms)b)$$

for arbitrary $s \in S$, $b \in B$ and $m \in M$. Since $m(sb) = (ms)b$, we have $\beta(sb) = s\beta(b)$. Therefore, β is an S-homomorphism.

Since we have the relations

$$\alpha(mb) = m\beta(b), \quad \beta(na) = n\alpha(a), \quad m \in M, \ n \in N, \ a \in A, \ b \in B,$$

the pair (α, β) is a K-homomorphism. The above relations follow from the property that $m\beta(b) = \beta(b)(m) = \alpha(mb)$. Next, we have

$$\beta(na)(m) = \alpha(m(na)) = \alpha((mn)a)) = mn\alpha(a) = (n\alpha(a))(m),$$
$$\beta(na) = n\alpha(a).$$

We assume that $(\alpha, \gamma) \colon (A, B) \to (X, H(X))$ is some K-homomorphism. By the definition of the module $(X, H(X))$, we have $\gamma(b)(m) = m\gamma(b)$, $b \in B$, $m \in M$. On the other hand, $m\gamma(b) = \alpha(mb)$ and $\beta(b)(m) = \alpha(mb)$, whence $\gamma(b)(m) = \beta(b)(m)$ and $\gamma = \beta$.

A similar proof holds for the modules (A, B) and $(T(Y), Y)$. □

Corollary 3.1.4 *For any R-module X, we have canonical ring isomorphisms*

$$\operatorname{End}_K(X, T(X)) \cong \operatorname{End}_R(X) \cong \operatorname{End}_K(X, H(X)).$$

A similar assertion holds for endomorphism rings of the modules Y, $(T(Y), Y)$ and $(H(Y), Y)$.

Assume that we have a K-module (X, Y) with homomorphisms of module multiplication

$$g \colon N \otimes_R X \to Y, \qquad f \colon M \otimes_S Y \to X.$$

We consider one more type of K-modules, namely, we define the K-module $(T(Y), T(X))$. We need to specify the corresponding homomorphisms of module multiplication, which we denote f' and g'. Set

$$f' = 1 \otimes f \colon N \otimes_R T(Y) \to T(X), \quad f'(n \otimes (m \otimes y)) = n \otimes my,$$
$$g' = 1 \otimes g \colon M \otimes_S T(X) \to T(Y), \quad g'(m \otimes (n \otimes x)) = m \otimes nx.$$

We adopt our earlier convention concerning the form of representation; we assume that nx' coincides with $f'(n \otimes x')$ and my' coincides with $g'(m \otimes y')$. We have the relations

$$m(n \otimes x) = g'(m \otimes (n \otimes x)) = m \otimes nx, \quad n(m \otimes y) = n \otimes my.$$

We have to verify the associativity relations, i.e., the relations

$$(m'n')x' = m'(n'x'), \quad (n'm')y' = n'(m'y')$$

for any elements $m' \in M$, $n' \in N$, $x' \in T(Y)$, $y' \in T(X)$.

Of course, we can assume that $x = m \otimes y$ and $y' = n \otimes x$ for some $m \in M$, $n \in N$, $x \in X$, $y \in Y$. By calculating, we obtain that the relations

$$m'(n'x') = m'(n'(m \otimes y)) = m'(n' \otimes my) = m' \otimes n'(my)$$
$$= m' \otimes (n'm)y = m'(n'm)y = (m'n')my = (m'n')m \otimes y,$$
$$(m'n')x' = (m'n')(m \otimes y) = (m'n')m \otimes y$$

hold. The second relation is similarly verified.

Thus, there exists a K-module $(T(Y), T(X))$ with homomorphisms of module multiplication f' and g'. The relations

$$n(m \otimes y) = n \otimes my, \qquad m(n \otimes x) = m \otimes nx$$

hold for all m, n, x, y. We easily verify that the mapping

$$(f, g) \colon (T(Y), T(X)) \to (X, Y)$$

is a K-homomorphism.

Remarks

(1). For any module (X, Y), we have the following four homomorphisms:

$$(1, g) \colon (X, T(X)) \to (X, Y), \quad (f, 1) \colon (T(Y), Y) \to (X, Y),$$
$$(1, f') \colon (X, Y) \to (X, H(X)), \quad (g', 1) \colon (X, Y) \to (H(Y), Y).$$

It follows from Lemmas 3.1.2 and 3.1.3 that f, g and f', g' are uniquely determined homomorphisms provided the second mapping is the identity mapping. These homomorphisms will be quite useful in what follows.

(2). We have a right-side variant of the constructions of the K-modules $(X, H(X))$ and $(H(Y), Y)$. Namely, if Z is a right R-module, then the group of row vectors $(Z, \mathrm{Hom}_R(N, Z))$ is a right K-module. The homomorphisms of module multiplication are defined similarly to the case of left modules. Similarly, a right S-module Z leads to the right K-module $(\mathrm{Hom}_S(M, Z), Z)$.

We point out some functors which act between the categories K-Mod, R-mod and S-Mod. For any formal matrix ring K, there exists a ring homomorphism

$$R \times S \to K, \qquad (r, s) \to \begin{pmatrix} r & 0 \\ 0 & s \end{pmatrix}.$$

Consequently, every K-module can be considered as an $R \times S$-module. This provides the "forgetful" functor $E: K\text{-Mod} \to R \times S\text{-mod}$. We remark that if K is a formal matrix ring with zero trace ideals, then the "diagonal" mapping

$$K \to R \times S, \qquad \begin{pmatrix} r & m \\ n & s \end{pmatrix} \to (r, s)$$

is a homomorphism; consequently, each $R \times S$-module is a K-module in this case. Now we define two functors T, H from $R \times S$-Mod into K-Mod. Given an $R \times S$-module (X, Y), the functors T, H put in line the K-modules $(X, T(X)) \oplus (T(Y), Y)$ and $(X, H(X)) \oplus (H(Y), Y)$, respectively. If $(\alpha, \beta): (X, Y) \to (X_1, Y_1)$ is an $R \times S$-module homomorphism, then $T(\alpha, \beta)$ and $H(\alpha, \beta)$ are the induced homomorphisms $T(X, Y) \to T(X_1, Y_1)$ and $H(X, Y) \to H(X_1, Y_1)$ defined in an obvious way. In fact, T is the functor $K \otimes_{R \times S} (-)$, where $T(X, Y) = K \otimes_{R \times S} (X, Y)$, H is the functor $\text{Hom}_{R \times S}(K, -)$, where $H(X, Y) = \text{Hom}_{R \times S}(K, (X, Y))$. The functors T and H transfer the $R \times S$-module homomorphisms into the corresponding induced K-module homomorphisms. The functor T (resp., the functor H) is left conjugate (resp., right conjugate) to the functor E. These conjugation situations appear in Lemmas 3.1.2 and 3.1.3.

The functors T and H are also related to each other via a natural transformation $\theta: T \to H$. The corresponding natural homomorphism $\theta(X, Y): T(X, Y) \to H(X, Y)$ is the sum of the homomorphisms $(1, h) + (h', (1)$, where $h: T(X) \to H(X)$ maps from $n \otimes x$ onto the homomorphism $m \to (mn)x$. We define $(h', (1)$: $(T(Y), Y) \to (H(Y), Y)$ similarly.

We define the functor $T_N = N \otimes_R (-): R\text{-Mod} \to S\text{-Mod}$ by the relation $T_N(X) = N \otimes_R X$ for any R-module X. The functor T_N maps the R-module homomorphisms to the induced S-module homomorphisms. The functor T_M is similarly determined.

Also, there exist two functors $(1, T_N): R\text{-Mod} \to K\text{-Mod}$ and $(1, 0): K\text{-Mod} \to R\text{-mod}$. The first functor is actually the restriction of the above functor T. Namely, $(1, T_N)X = (X, T_N(X))$ and $(1, 0)(X, Y) = X$ for any R-module X and every K-module (X, Y). The both functors transfer homomorphisms to the induced homomorphisms. We have the relations

$$((1, 0)(1, T_N))X = X \quad \text{and} \quad ((1, T_N)(1, 0))(X, Y) = (X, T_N(X)).$$

There are also similar functors $(T_M, 1): S\text{-Mod} \to K\text{-Mod}$ and $(0, 1): K\text{-Mod} \to S\text{-mod}$.

Now, we define the functor (T, T) from the category T-Mod to itself. The functor (T, T) maps a module (X, Y) onto the module $(T(Y), T(X))$ and the homomorphism $(\alpha, \beta): (X, Y) \to (X', Y')$ is mapped onto the homomorphism $(T(\beta), T(\alpha)): (T(Y), T(X)) \to (T(Y'), T(X'))$. We verify that the mapping $(T(\beta), T(\alpha))$ is a K-homomorphism as follows. With the use of the relations $\alpha(my) = m\beta(y)$ and $\beta(nx) = n\alpha(x)$, we obtain the relations $T(\beta)(my') = mT(\alpha)(y')$ and $T(\alpha)(nx') = nT(\beta)(x')$, where $y' = n \otimes x$ and $x' = m \otimes y$. Therefore,

$$T(\beta)(my') = T(\beta)(m(n \otimes x)) = T(\beta)(m \otimes nx) = m \otimes \beta(nx)$$
$$= m \otimes n\alpha(x) = m(n \otimes \alpha(x)) = m((1 \otimes \alpha)(n \otimes x)) = mT(\alpha)(y').$$

The second relation is obtained similarly.

We now consider the form of submodules and factor modules of K-modules. This is easy to do, since we know the structure of K-modules. Assume that we have a module $V = (X, Y)$ over the ring K. A subset $W \subseteq V$ is a submodule of the module V if and only if there exist a submodule A of the R-module X and a submodule B of the S-module Y such that $W = (A, B)$, $MB \subseteq A$ and $NA \subseteq B$. The following is an important special case. For submodules A and B of the modules X and Y, respectively, the sets (A, NA) and (MB, B) are submodules of (X, Y). If the ring K has zero trace ideals (i.e., $I = 0 = J$), then we obtain the submodules $(MB, 0)$ and $(0, NA)$. The interrelation between the modules (A, NA), (MB, B) and the modules $(A, T(A))$, $(T(B), B)$ and $(A, H(A))$, $(H(B), B)$ is clear from the remarks after Corollary 3.1.4.

Let $W = (A, B)$ be a submodule of a K-module $V = (X, Y)$. The group of row vectors $(X/A, Y/B)$ is a K-module. The homomorphisms of module multiplication

$$M \otimes_S Y/B \to X/A, \qquad N \otimes_R X/A \to Y/B$$

are induced by the homomorphisms of module multiplication in the module (X, Y). Namely, $m\overline{y} = \overline{my}$, where $\overline{y} = y + B$, $\overline{my} = my + A$, and we have a similar situation for the second homomorphism. The factor module V/W can be identified with the module $(X/A, Y/B)$. More precisely, the correspondence $(x, y) + W \to (x + A, y + B)$ is an isomorphism between these modules.

When working with modules over the ring of formal triangular matrices, some special features appear. It is easy to find them by considering the previous text under the condition $N = 0$. We will only focus on a few details. Let (X, Y) be a K-module. In this case, the homomorphism g of module multiplication is equal to the zero homomorphism. The two associativity relations $(*)$ from Sect. 2.1 obviously hold. An important feature of the "triangular case" is that for any R-module X, we have the K-module $(X, 0)$. A K-module homomorphism $(X, Y) \to (X_1, Y_1)$ is a pair (α, β) consisting of an R-homomorphism $\alpha \colon X \to X_1$ and an S-homomorphism $\beta \colon Y \to Y_1$ which satisfy the relation $\alpha(my) = m\beta(y)$ for all $m \in M$ and $y \in Y$. The category $\mathcal{A}(K)$ from Theorem 3.1.1 is turned into the category of "triples" of the form (X, Y, f). Diagrams (3.1.1) are always commutative, and only the first diagram remains in (3.1.2). If X is an R-module and Y is an S-module, then the K-modules $(X, T(X))$ and $(H(Y), Y)$ from Lemmas 3.1.2 and 3.1.3 have the form $(X, 0)$ and $(0, Y)$, respectively.

The structure of modules over formal matrix rings of order $n > 2$ is similar to the structure of modules over formal matrix rings of order $n = 2$. Such modules are modules of column vectors of height n, and the module multiplication acts by the rule "multiplication of the matrix by column". There is no need to give full details since everything is clear from the considered case $n = 2$. We consider modules over the ring of formal triangular matrices

$$\Gamma = \begin{pmatrix} R & M & L \\ 0 & S & N \\ 0 & 0 & T \end{pmatrix}$$

of order 3 in more detail; see the end of Sect. 2.3 for more about this ring. Let $\varphi: M \otimes_S N \to L$ be an R-T-bimodule homomorphism from the definition of the ring Γ. As earlier, we write mn instead of $\varphi(m \otimes n)$. We assume that there are an R-module X, an S-module Y, a T-module Z, R-module homomorphisms $f: M \otimes_S Y \to X$ and $h: L \otimes_T Z \to X$, an S-module homomorphism $g: N \otimes_T Z \to Y$ (the previous notations are preserved), and $m(nz) = (mn)z$ for all $m \in M, n \in N, z \in Z$. Then (X, Y, Z) is a Γ-module with module multiplication of "a matrix by a column" type. Via this method, we can obtain each Γ-module.

The homomorphisms of Γ-modules act coordinate-wise. A homomorphism $(X, Y, Z) \to (X_1, Y_1, Z_1)$ is a triple (α, β, γ), where $\alpha: X \to X_1$, $\beta: Y \to Y_1$, $\gamma: Z \to Z_1$ are homomorphisms of the corresponding modules. In addition, the relations

$$\alpha(my) = m\beta(y), \quad \alpha(\ell z) = \ell\gamma(z), \quad \beta(nz) = n\gamma(z)$$

have to hold for all values m, n, ℓ, y, z.

Similar to the category $\mathcal{A}(K)$, we define the category $\mathcal{A}(\Gamma)$. In this category, the objects are expressions (X, Y, Z, f, g, h), the morphisms are triples (α, β, γ) such that the diagrams similar to diagrams (2) before Theorem 3.1.1 are commutative. The categories Γ-Mod and $\mathcal{A}(\Gamma)$ are equivalent [42]. This can be proved by direct arguments, which are similar to those in the proof of Theorem 3.1.1. We can also apply Theorem 3.1.1 twice. We detail the second assertion.

The ring Γ is naturally isomorphic to both rings Δ and Λ of triangular matrices of order 2; see the end of Sect. 2.3. In turn, a Γ-module (X, Y, Z) with homomorphisms f, g, h can be considered as a Δ-module $((X, Y), Z)$ of row vectors of length 2, consisting of blocks (x, y) and z. Here (X, Y) is an $\begin{pmatrix} R & M \\ 0 & S \end{pmatrix}$-module, which is obtained with the use of the homomorphism $f: M \otimes_S Y \to X$, and $h + g$ is the homomorphism $\begin{pmatrix} L \\ N \end{pmatrix} \otimes_T Z \to (X, Y)$. The module multiplication of the Γ-module (X, Y, Z) induces the module multiplication of the Δ-module $((X, Y), Z)$ performed over blocks. Similarly, the Γ-module (X, Y, Z) is turned into a Λ-module $(X, (Y, Z))$ in such a way that (Y, Z) is an $\begin{pmatrix} S & N \\ 0 & T \end{pmatrix}$-module, which is obtained with the use of the homomorphism $g: N \otimes_T Z \to Y$ and the homomorphism $(M, L) \otimes_{\begin{pmatrix} S & N \\ 0 & T \end{pmatrix}} (Y, Z) \to X$ is $f + h$ (we need to make use of the fact that $m(nz) = (mn)z$). There is no practical difference between the Γ-module (X, Y, Z)

and the Δ-module $((X, Y), Z)$. More precisely, we can say that the categories Γ-Mod and Δ-Mod are equivalent. Now it is clear how we can obtain the equivalence of the categories Γ-Mod and $\mathcal{A}(\Gamma)$ with a double application of Theorem 3.1.1.

Results in the other direction are also interesting and important. We mean the reduction of the study modules over an arbitrary ring $\begin{pmatrix} R & M \\ N & S \end{pmatrix}$ to the study modules over some ring of triangular matrices. Here is one simple theorem of a quite general type in this field.

Let $K = \begin{pmatrix} R & M \\ N & S \end{pmatrix}$ be a formal matrix ring with bimodule homomorphisms $\varphi \colon M \otimes_S N \to R$ and $\psi \colon N \otimes_R M \to S$. We fix some ideal L of the ring R containing the trace ideal I (for example, $L = I$ or $L = R$). Then there exists a ring of triangular matrices

$$\begin{pmatrix} R & M & L \\ 0 & S & N \\ 0 & 0 & R \end{pmatrix}$$

of order 3. Such rings have just been considered. We take φ as a bimodule homomorphism $M \otimes_S N \to L$; this is well defined, since $\operatorname{Im} \varphi = I \subseteq L$.

Let $V = (X, Y)$ be some K-module with homomorphisms of module multiplication $f \colon M \otimes_S Y \to X$ and $g \colon N \otimes_R X \to Y$. We can construct a Γ-module $W = (X, Y, X)$ with homomorphisms of module multiplication $f \colon M \otimes_S Y \to X$, $h \colon L \otimes_R X \to X$ and $g \colon N \otimes_R X \to Y$, where h is the canonical homomorphism $\ell \otimes x \to \ell x$, $\ell \in L$, $x \in X$. The relation $m(nz) = (mn)z$ holds, since it is transformed into the relation $m(nx) = (mn)x$, which follows from the existence of the K-module (X, Y).

We find interrelations between the K-module homomorphisms and the Γ-module homomorphisms. Let $(\alpha, \beta) \colon (X, Y) \to (X_1, Y_1)$ be a K-module homomorphism. We assert that $(\alpha, \beta, \alpha) \colon (X, Y, X) \to (X_1, Y_1, X_1)$ is a Γ-module homomorphism. Indeed, the three required relations

$$\alpha(my) = m\beta(y), \quad \beta(nx) = n\alpha(x), \quad \alpha(\ell x) = \ell\alpha(x)$$

hold. Conversely, we assume that $(\alpha, \beta, \gamma) \colon (X, Y, X) \to (X_1, Y_1, X_1)$ is a Γ-module homomorphism and we assume that $L = R$. Then

$$r\alpha(x) = \alpha(rx) = r\gamma(x) \quad \text{for all} \quad r \in R, x \in X.$$

Therefore, we have $\gamma = \alpha$. Thus, every homomorphism $(X, Y, X) \to (X_1, Y_1, X_1)$ is "the triple" (α, β, α).

We represent the obtained interrelation between K-modules and Γ-modules in a category form. We take the ring

$$\begin{pmatrix} R & M & R \\ 0 & S & N \\ 0 & 0 & R \end{pmatrix}$$

as Γ. We define a covariant functor $F \colon K\text{-Mod} \to \Gamma\text{-mod}$. The functor F transfers a K-module $V = (X, Y)$ onto the Γ-module $F(V) = (X, Y, X)$. The functor F transfers a K-module homomorphism (α, β) onto the Γ-module homomorphism (α, β, α).

Theorem 3.1.5 *The functor F provides a full embedding of the category K-Mod into the category Γ-mod.*

Proof We mean that F determines the equivalence between K-Mod and a full subcategory in the category Γ-Mod consisting of the modules of the form (X, Y, X). Everything we need to prove this assertion is given in the text before Theorem 3.1.5. $\square$

Corollary 3.1.6 *Let V be a K-module.*

(1) *The endomorphism rings of the K-module V and the Γ-module $F(V)$ are isomorphic.*
(2) *The K-module V is indecomposable if and only if the Γ-module $F(V)$ is indecomposable.*

In [58], representations of Γ-modules are studied in detail, where Γ is an arbitrary ring of triangular matrices

$$\begin{pmatrix} R & M & L \\ 0 & S & N \\ 0 & 0 & T \end{pmatrix};$$

Γ-modules may also be considered as modules over each of two rings of triangular matrices of order 2 considered in Sect. 2.3. The paper [42] contains an interesting study of the problem mentioned in Theorem 3.1.5. In particular, the author constructs several functors from the category K-modules into the category of modules over various rings of triangular matrices of order 3. The author tries to choose Γ so that K and Γ are both of finite (infinite) representation type.

The modules over a formal matrix ring of order $n > 2$ are arranged like the modules in the case $n = 2$. Such modules are modules of column vectors of height n and the module multiplication is the multiplication of the matrix by a column. A concrete K-module X has the form $\begin{pmatrix} X_1 \\ X_2 \\ \cdots \\ X_n \end{pmatrix}$, where X_i is an R_i-module, $i = 1, \ldots, n$. In addition, for any k, there exist homomorphisms $f_{ki} \colon M_{ki} \otimes_{R_i} X_i \to X_k$, $i = 1, \ldots, n$, and the obvious associativity relations hold. Set $f_{ki}(a \otimes x) = ax$, where $a \in M_{ki}$ and $x \in X_i$. Usually, we represent the module X and elements of X as rows.

All functors defined for $n = 2$ have analogues for $n > 2$. We define only the analogue of the functor T. We denote by L the ring direct product $R_1 \times R_2 \times \ldots \times R_n$. The diagonal embedding $i\colon L \to K$ induces the functor $T = T(i) = K \otimes_L (-)\colon L\text{-Mod} \to K\text{-mod}$. If $X = (X_1, \ldots, X_n)$ is an L-module, then

$$T(X) = (X_1, T(X_1), \ldots, T(X_1)) \oplus (T(X_2), X_2, \ldots, T(X_2)) \oplus \ldots$$
$$\oplus (T(X_n), T(X_n), \ldots, X_n).$$

Employing an abbreviated notation, we have $T(X) = (X_1, T(X_1)) \oplus \ldots \oplus (T(X_n), X_n)$; see [33].

3.2 Small and Essential Submodules

We demonstrate some ways of working with modules over formal matrix rings. Let $K = \begin{pmatrix} R & M \\ N & S \end{pmatrix}$ be a formal matrix ring and (X, Y) a K-module. Some submodules of the module (X, Y) play a very important role in various questions. In particular, the submodules MY and NX, defined in Sect. 3.1, are such submodules. We also define two submodules by setting $L(X) = \{x \in X \mid nx = 0 \text{ for every } n \in N\}$ and $L(Y) = \{y \in Y \mid my = 0 \text{ for every } m \in M\}$. If the ring K has zero trace ideals, then $MY \subseteq L(X)$ and $NX \subseteq L(Y)$.

In this section, we assume that K is a ring with zero trace ideals, i.e., the trace ideals I and J of the ring K are equal to the zero ideal. In this case, $mn = 0 = nm$ for all $m \in M$ and $n \in N$. In addition to small and essential submodules, we describe finitely generated, hollow, and uniform K-modules. It is appropriate to make a remark similar to that made in Sect. 2.1 concerning the nature of research on the ring K.

When we speak about a description of some K-module (X, Y), it is reasonable to look for a description in terms of the R-module X, the S-module Y, and the actions of the bimodules M and N on Y and X, respectively.

Proposition 3.2.1 *A K-module (X, Y) is finitely generated if and only if the R-module X/MY and the S-module Y/NX are finitely generated.*

Proof Let (X, Y) be a finitely generated K-module with a finite generator system $(x_1, y_1), \ldots, (x_k, y_k)$. We take an arbitrary element $x \in X$ and write $(x, 0) = t_1(x_1, y_1) + \ldots + t_k(x_k, y_k)$, where $t_i = \begin{pmatrix} r_i & m_i \\ n_i & s_i \end{pmatrix}$, $i = 1, \ldots, k$. Then

$$x = \sum_{i=1}^{k}(r_i x_i + m_i y_i), \quad x + MY = r_1(x_1 + MY) + \ldots + r_k(x_k + MY).$$

Consequently, $\{x_i + MY\}_{i=1}^{k}$ and $\{y_i + NX\}_{i=1}^{k}$ are generator systems for X/MY and Y/NX, respectively.

We now assume that the modules X/MY, Y/NX are finitely generated and $\{x_i + MY\}_{i=1}^k$, $\{y_i + NX\}_{i=1}^\ell$ are their generator systems. We assert that $\{(x_i, 0), (0, y_j) \mid i = 1, \ldots, k, \; j = 1, \ldots, \ell\}$ is a generator system of the K-module (X, Y). It is sufficient to verify that all elements of the form $(x, 0)$ and $(0, y)$ are linear combinations of elements of the above system. For $(x, 0)$, this is obtained as follows (for $(0, y)$, we use a similar argument). We have $x = r_1 x_1 + \ldots + r_k x_k + m_1 b_1 + \ldots + m_i b_i$, where $r \in R$, $m \in M$, $b \in Y$. In turn, each of the elements $b_1, \ldots, b_i$ is equal to a sum of the form $s_1 y_1 + \ldots + s_\ell y_\ell + n_1 a_1 + \ldots + n_j a_j$, where $s \in S, n \in N, a \in X$. We substitute these sums into the expression for x bearing in mind that we always have $mn = 0$. We obtain that $x = r_1 x_1 + \ldots + r_k x_k + c_1 y_1 + \ldots + c_\ell y_\ell$ for some $c_1, \ldots, c_\ell \in M$. Now it is clear how to represent $(x, 0)$ in the required form. We remark only that, for example, $\begin{pmatrix} 0 & c_1 \\ 0 & 0 \end{pmatrix}(0, y_1) = (c_1 y_1, 0)$. $\qquad\square$

We denote by (σ, τ) the canonical homomorphism $(X, Y) \to (X/MY, Y/NX)$. We see that (MY, NX) is a submodule of (X, Y) and the factor module $(X, Y)/(MY, NX)$ can be identified with the module $(X/MY, Y/NX)$, as we agreed in Sect. 3.1.

A submodule A of some module V is said to be *small* if $B = V$ for any submodule B in V with $A + B = V$.

Proposition 3.2.2 *Let (X, Y) be a K-module. A submodule (A, B) is small in (X, Y) if and only if σA is a small submodule of X/MY and τB is a small submodule of Y/NX.*

Proof We assume that (A, B) is a small submodule of (X, Y). Assume that we have the relation $\sigma A + C/MY = X/MY$ for some submodule C in X. Since $\sigma A = (A + MY)/MY$, we have

$$A + C = X, \quad (A, B) + (C, Y) = (X, Y), \quad (C, Y) = (X, Y), \quad C = X.$$

Therefore, σA is a small submodule of X/MY. Similarly, one can prove that τB is a small submodule of Y/NX.

Now let us assume that σA is a small submodule of X/MY and τB is a small submodule of Y/NX. Let $(A, B) + (C, D) = (X, Y)$ for some submodule (C, D) in (X, Y). Then

$$A + C = X, \quad B + D = Y, \quad \sigma A + (C + MY)/MY = X/MY,$$
$$\tau B + (D + NX)/NX = Y/NX.$$

Since σA is a small submodule of X/MY and τB is a small submodule of Y/NX, we have that $C + MY = X$ and $D + NX = Y$. We multiply the last relations by N and M, respectively. We have $NC = NX$ and $MD = MY$. Now we obtain $MY = MD \subseteq C$ and $NX = NC \subseteq D$. Therefore, $C = X$, $D = Y$ and (A, B) is a small submodule of (X, Y). $\qquad\square$

A module M is said to be *hollow* if M is not equal to the zero module and all its submodules are small in M.

Corollary 3.2.3 *A non-zero module (X, Y) is hollow if and only if either X/MY is a hollow module and $Y = NX$, or Y/NX is a hollow module and $X = MY$.*

Proof First, we remark that the relations $X = MY$ and $Y = NX$ cannot both be true.

We assume that (X, Y) is a hollow module. We consider two possible cases for the module (X, NX).

(1). $(X, NX) = (X, Y)$. Then $Y = NX$ and $X \neq MY$. We take some submodule A/MY in X/MY, where $A \neq X$. By assumption, (A, NA) is a small submodule of (X, Y). It follows from Proposition 3.2.2 that A/MY is a small submodule of X/MY. Therefore, X/MY is a hollow module.

(2). $(X, NX) \neq (X, Y)$. It follows from Proposition 3.2.2 that X/MY is a small submodule of X/MY, whence $X = MY$. Then $(MY, Y) = (X, Y)$ and, similar to Item (1), we obtain that Y/NX is a hollow module.

Now assume that X/MY is a hollow module and $Y = NX$. (The case where Y/NX is a hollow module and $X = MY$ is considered similarly.) We take some proper submodule (A, B) in (X, Y). It is clear that $A \neq X$. By assumption, σA is a small submodule of X/MY, and it is obvious that τB is a small submodule of Y/NX. By Proposition 3.2.2, (A, B) is a small submodule of (X, Y). Therefore, (X, Y) is a hollow module.   □

A module is said to be *local* if it is finitely generated and has exactly one maximal submodule. It is not difficult to verify that local modules coincide with finitely generated hollow modules. The next result follows from Proposition 3.2.1 and Corollary 3.2.3.

Corollary 3.2.4 *A module (X, Y) is local if and only if either X/MY is a local module and $Y = NX$, or Y/NX is a local module and $X = MY$.*

A submodule A of the module V is said to be *essential* or *large* if A has non-zero intersection with every non-zero submodule of the module V. In this case, V is called an *essential extension* of the module A.

Proposition 3.2.5 *A submodule (A, B) of the module (X, Y) is essential if and only if $A \cap L(X)$ is an essential submodule of $L(X)$ and $B \cap L(Y)$ is an essential submodule of $L(Y)$.*

Proof Let (A, B) be an essential submodule of (X, Y). We verify that $A \cap L(X)$ is an essential submodule of $L(X)$. For a non-zero element $x \in L(X)$, there exists a matrix $\begin{pmatrix} r & m \\ n & s \end{pmatrix}$ such that

$$0 \neq \begin{pmatrix} r & m \\ n & s \end{pmatrix}(x, 0) \in (A, B) \quad \text{or} \quad (rx, nx) \in (A, B).$$

In addition, $nx = 0$, since $x \in L(X)$. Then $rx \neq 0$, $rx \in A \cap L(X)$, and $A \cap L(X)$ is an essential submodule of $L(X)$. Similarly, $B \cap L(Y)$ is an essential submodule of $L(Y)$.

We assume now that $A \cap L(X)$ is an essential submodule of $L(X)$, $B \cap L(Y)$ is an essential submodule of $L(Y)$ and (x, y) is a non-zero element in (X, Y). We prove that $K(x, y) \cap (A, B) \neq 0$. If $y \neq 0$, then we can assume that $x = 0$. We consider the case where $y \notin L(Y)$. Then $my \neq 0$ for some $m \in M$. Since $my \in L(X)$ and $A \cap L(X)$ is an essential submodule of $L(X)$, we have that $0 \neq rmy \in A \cap L(X)$ for some $r \in R$. Then

$$\begin{pmatrix} 0 & rm \\ 0 & 0 \end{pmatrix} (0, y) = (rmy, 0) \in (A, B).$$

If $y \in L(Y)$, then $0 \neq sy \in B \cap L(Y)$, where $s \in S$, since $B \cap L(Y)$ is an essential submodule of $L(Y)$. Then $s(0, y) = (0, sy) \in (A, B)$. The case $y = 0$ is considered similarly. □

A module is said to be *uniform* if the intersection of any two of its non-zero submodules is not equal to the zero module.

Corollary 3.2.6 *A module (X, Y) is uniform if and only if either $L(X) = 0$ and Y is uniform, or $L(Y) = 0$ and X is uniform.*

Proof Let (X, Y) be a uniform module. The intersection of the submodules $(L(X), 0)$ and $(0, L(Y))$ is equal to the zero module. Therefore, at least one of the modules $L(X), L(Y)$ is equal to the zero module. If, for example, $L(X) = 0$, then $L(Y) = Y$, since $MY \subseteq L(X) = 0$. We take an arbitrary non-zero submodule B in Y. Since (MB, B) is an essential submodule of (X, Y), it follows from Proposition 3.2.5 that $B \cap L(Y)$ is an essential submodule of $L(Y)$. Therefore, Y is an essential extension of the module B and the module Y is uniform. If $L(Y) = 0$, then we argue similarly.

For the proof of the converse assertion, we assume that $L(X) = 0$ and Y is uniform. As above, $L(Y) = Y$. Let (A, B) be an arbitrary non-zero submodule of (X, Y). Then $B \neq 0$, since otherwise $A \neq 0$ and $A \subseteq L(X)$, which is impossible. It follows from Proposition 3.2.5 that (A, B) is an essential submodule of (X, Y). Therefore, the module (X, Y) is uniform. □

Remark All the results of this section can be applied to modules over the ring $\begin{pmatrix} R & M \\ 0 & S \end{pmatrix}$ of triangular matrices (see [52]). We only have to add some details, keeping in mind the property that for a module (X, Y) over such a ring, the relations $NX = 0$ and $L(X) = X$ always hold. For example,

(X, Y) is a hollow module if and only if either X is a hollow module and $Y = 0$, or Y is a hollow module and $X = MY$;

(X, Y) is a uniform module if and only if either $X = 0$ and Y is a uniform module, or $L(Y) = 0$ and X is a uniform module.

3.3 The Socle and the Radical

In this section, K is an arbitrary formal matrix ring $\begin{pmatrix} R & M \\ N & S \end{pmatrix}$. First, we describe simple K-modules. Then we use this description to study the structure of minimal submodules, maximal submodules, the socle, and the radical.

A non-zero module is said to be *simple* if it does not have non-trivial submodules.

Proposition 3.3.1 *A module* (X, Y) *is simple if and only if either* X, Y *are simple modules and* $X = MY, Y = NX$, *or* X *is a simple module and* $Y = 0$, *or* $X = 0$ *and* Y *is a simple module.*

Before proving Proposition 3.3.1, we make the following remark.

Since the inclusions $IX \subseteq MY$ and $JY \subseteq NX$ always hold (see Sect. 3.1), it follows from the relations $X = IX$ and $Y = JY$ that $X = MY$ and $Y = NX$. We also have the converse. Similarly, the condition $IX, JY \neq 0$ is equivalent to the condition $MY, NX \neq 0$. Therefore, in Proposition 3.3.1, we can write $X = IX$, $Y = JY$ or $IX, JY \neq 0$, or $MY, NX \neq 0$.

We pass to the proof of Proposition 3.3.1. Let (X, Y) be a simple module and $X \neq 0$, $Y \neq 0$. For any non-zero submodules A in X and B in Y, we have that (A, NA) and (MB, B) are submodules of (X, Y). Therefore, $A = X$, $B = Y$ and X, Y are simple modules. In particular, $X = MY$ and $Y = NX$. If one of the modules X, Y is equal to the zero module, then it is clear that the second module is simple.

Now we assume that X, Y are simple modules and $X = MY, Y = NX$. Any non-trivial submodule of (X, Y) is necessarily of the form $(X, 0)$ or $(0, Y)$. This is impossible, since $X = MY$ and $Y = NX$. If one of the modules X, Y is simple and the second module is equal to the zero module, then it is clear that (X, Y) is a simple module.   □

Corollary 3.3.2 *Let* (X, Y) *be a* K-module.

(1) *If* $L(X) = 0 = L(Y)$, *then the module* (X, Y) *is simple if and only if* X *and* Y *are simple modules.*

(2) *If* K *is a ring with zero trace ideals, then the module* (X, Y) *is simple if and only if either the module* X *is simple and* $Y = 0$, *or* $X = 0$ *and the module* Y *is simple.*

A non-zero (resp., proper) submodule of a module V is said to be *minimal* (resp., *maximal*) if it is a minimal (resp., maximal) element in the lattice of all submodules of the module V.

Corollary 3.3.3 *Let* (A, B) *be a submodule of a* K-module (X, Y).

(1) (A, B) *is minimal if and only if either* A, B *are minimal and* $A = MB, B = NA$, *or* A *is minimal and* $B = 0$ *(then* $NA = 0$*), or* $A = 0$ *(then* $MB = 0$*) and* B *is minimal.*

(2) (A, B) *is maximal if and only if either* A, B *are maximal and* $MY \nsubseteq A$, $NX \nsubseteq B$ *(this is equivalent to the property that* $IX \nsubseteq A$ *and* $JY \nsubseteq B$), *or* A *is maximal and* $B = Y$ *(then* $MY \subseteq A$), *or* $A = X$ *(then* $NX \subseteq B$) *and* B *is maximal.*

Proof The assertion (1) directly follows from Proposition 3.3.1, since every minimal submodule is a simple module. In relation to assertion (2), we remark that (A, B) is a maximal submodule if and only if $(X, Y)/(A, B) = (X/A, Y/B)$ is a simple module. Again, we can use Proposition 3.3.1. If the first possibility from Proposition 3.3.1 holds, then $X/A = M(Y/B)$ and $Y/B = N(X/A)$. Since $M(Y/B) = (MY + A)/A$, we have that $X = MY + A$ and $MY \nsubseteq A$, since A is maximal. In addition, $NX \nsubseteq B$. Similarly, one can prove that $IX \nsubseteq A$ and $JY \nsubseteq B$; this also follows from the remark after Proposition 3.3.1. □

The sum of all minimal submodules of the module V is called the *socle* of V; it is denoted by Soc V. If V does not have any minimal submodules, then Soc $V = 0$ by definition.

Corollary 3.3.4 *Let* (X, Y) *be a module. The socle of* (X, Y) *is equal to* (Soc $L(X)$, Soc $L(Y)$) $+ \sum(A, NA)$, *where the summation is over all minimal submodules* A *in* X *such that* $IA \neq 0$ *and* NA *is a minimal submodule of* B. *The last summand is also equal to* $\sum(MB, B)$, *where the summation is over all minimal submodules* B *in* Y *such that* $JB \neq 0$ *and* MB *is a minimal submodule of* A; *this summand is also equal to* $\sum(A, B)$, *where the summation is over all above-mentioned* A *and* B.

Proof Since there are three types of minimal submodules, we can represent three sums of the corresponding minimal submodules and obtain

$$\text{Soc}(X, Y) = \sum(A, 0) + \sum(0, B) + \sum(A, B).$$

Soc $L(X)$ is the first summand, Soc $L(Y)$ is the second summand, and the third summand coincides with each of the three sums mentioned in the corollary. □

Corollary 3.3.5 *Let* (X, Y) *be a* K-*module.*

(1) *If* $L(X) = 0 = L(Y)$, *then* $\text{Soc}(X, Y) = (\text{Soc } X, \text{Soc } Y)$.
(2) *If* K *is a ring with zero trace ideals, then* $\text{Soc}(X, Y) = (\text{Soc } L(X), \text{Soc } L(Y))$, *and* (X, Y) *is an essential extension of* $\text{Soc}(X, Y)$ *if and only if* X *is an essential extension of* $\text{Soc } L(X)$ *and* Y *is an essential extension of* $\text{Soc } L(Y)$.

The intersection of all maximal submodules of the module V is called the *radical* of the module V; it is denoted by Rad V. If V does not have maximal submodules, then Rad $V = V$ by definition.

Below, (σ, τ) denotes the canonical homomorphism $(X, Y) \to (X/MY, Y/NX)$.

Corollary 3.3.6 *The radical of the module* (X, Y) *is equal to* $(\sigma^{-1}(\text{Rad } X/MY), \tau^{-1}(\text{Rad } Y/NX)) \cap (\cap(A, B))$, *where the intersection is over all submodules* (A, B) *in* (X, Y) *such that* A, B *are maximal submodules of* X, Y, *respectively, and* $MY \nsubseteq A$, $NX \nsubseteq B$.

Proof We know that maximal submodules of (X, Y) can be of three types. Therefore, we can consider three families of the corresponding maximal submodules and represent

$$\mathrm{Rad}(X, Y) = (\cap(A, Y)) \cap (\cap(X, B)) \cap (\cap(A, B)),$$

$$\cap(A, Y) = (\sigma^{-1}(\mathrm{Rad}\, X/MY), Y), \quad \cap(X, B) = (X, \tau^{-1}(\mathrm{Rad}\, Y/NX)). \quad \square$$

Corollary 3.3.7 *Let (X, Y) be a K-module.*

(1) *If $NX = Y$ and $MY = X$, then $\mathrm{Rad}(X, Y) = (\mathrm{Rad}\, X, \mathrm{Rad}\, Y)$.*
(2) *If K is a ring with zero trace ideals, then $\mathrm{Rad}(X, Y) = (\sigma^{-1}(\mathrm{Rad}\, X/MY), \tau^{-1}$ $(\mathrm{Rad}\, Y/NX))$, and $\mathrm{Rad}(X, Y)$ is a small submodule of (X, Y) if and only if $\mathrm{Rad}\, X/MY$ is a small submodule of X/MY and $\mathrm{Rad}\, Y/NX$ is a small submodule of Y/NX.*

Proof (1). It follows from Corollary 3.3.6 that $\mathrm{Rad}(X, Y) = \cap(A, B)$, where (A, B) are submodules of (X, Y) such that A and B are maximal submodules. It is not obvious that this intersection coincides with $(\mathrm{Rad}\, X, \mathrm{Rad}\, Y)$. Since $\cap(A, B) = (\cap A, \cap B)$, it is sufficient to prove the following two properties:

(I) for every maximal submodule A in X, there exists a maximal submodule B in Y such that (A, B) is a submodule of (X, Y) (then (A, B) is maximal);
(II) for every maximal submodule B in Y, there exists a maximal submodule A in X such that (A, B) is a submodule of (X, Y) (then (A, B) is maximal).

Let A be some maximal submodule of X. The subset (A, Y) is not a submodule, since $MY = X$. Therefore, the set of all submodules of the form (A, D) is inductive and not empty (since it contains the submodule (A, NA)). By Zorn's lemma, this set contains a maximal element (A, B). This submodule is a maximal submodule of (X, Y) (we take into account the relation $NX = Y$). By Corollary 3.3.3, B is a maximal submodule of Y, which is what we required. For a maximal submodule B in Y, a similar argument can be used.

(2). If (A, B) is a submodule of (X, Y) such that A, B are maximal and $MY \not\subseteq A$, $NX \not\subseteq B$, then $IX \not\subseteq A$ and $JY \not\subseteq B$. Since $I = 0 = J$, the intersection $\cap(A, B)$ from Corollary 3.3.6 is equal to the zero. The second assertion of (2) follows from Proposition 3.2.2. $\square$

Remark All the results of this section can be applied to modules over the ring $\begin{pmatrix} R & M \\ 0 & S \end{pmatrix}$ of triangular matrices; see also the end of the previous section. The corresponding results are obtained in [52]. For example,

$$\mathrm{Soc}(X, Y) = (\mathrm{Soc}\, X, \mathrm{Soc}\, L(Y)), \quad \mathrm{Rad}(X, Y) = (\sigma^{-1}(\mathrm{Rad}\, X/MY), \mathrm{Rad}\, Y).$$

3.4 Injective Modules and Injective Hulls

We determine the structure of injective modules over formal matrix rings. In this section, K is an arbitrary formal matrix ring $\begin{pmatrix} R & M \\ N & S \end{pmatrix}$. In this section and the following two sections, we usually deal with a K-module, denoted by (A, B).

Let X be an R-module and Y an S-module. Considered before Lemma 3.1.3, the K-modules $(X, \mathrm{Hom}_R(M, X))$ and $(\mathrm{Hom}_S(N, Y), Y)$ will play a very important role. We denote these modules by $(X, H(X))$ and $(H(Y), Y)$, respectively. We will constantly use Lemma 3.1.3. For example, the following result is a corollary to this lemma.

Proposition 3.4.1 $(X, H(X))$ *is an injective K-module if and only if X is an injective R-module. A similar assertion holds for the S-module Y and the K-module $(H(Y), Y)$.*

Proof Let X be an injective R-module. We assume that there is a K-module monomorphism $(i, j)\colon (A, B) \to (C, D)$ and a K-module homomorphism $(\alpha, \beta)\colon (A, B) \to (X, H(X))$. Since the module X is injective, there exists a homomorphism $\gamma\colon C \to X$ with $i\gamma = \alpha$. By Lemma 3.1.3, there exists a $\delta\colon D \to H(X)$ for which $(\gamma, \delta)\colon (C, D) \to (X, H(X))$ is a K-module homomorphism. By Lemma 3.1.3, $j\delta = \beta$. Consequently, $(i, j)(\gamma, \delta) = (\alpha, \beta)$ and the module $(X, H(X))$ is injective.

Conversely, let $(X, H(X))$ be an injective module. We assume that $i\colon A \to C$ is a monomorphism and $\alpha\colon A \to X$ is an R-module homomorphism. It follows from Lemma 3.1.3 that there are K-module homomorphisms

$$(i, j)\colon (A, H(A)) \to (C, H(C)), \quad (\alpha, \beta)\colon (A, H(A)) \to (X, H(X)).$$

We prove that j is a monomorphism. Indeed, if $j(\eta) = 0$, where $\eta \in H(A)$, then $j(\eta)(m) = i(m\eta) = i(\eta(m)) = 0$ for any $m \in M$, whence $\eta = 0$. Therefore, (i, j) is a monomorphism. Since the module $(X, H(X))$ is injective, there exists a homomorphism $(\gamma, \delta)\colon (C, H(C)) \to (X, H(X))$ with $(i, j)(\gamma, \delta) = (\alpha, \beta)$. Consequently, $i\gamma = \alpha$ and the module X is injective.

In the case of the module $(H(Y), Y)$, we argue similarly. $\qquad\square$

Remark It follows from Proposition 3.4.1 that the K-module $(X, 0)$ is injective if and only if X is an injective R-module and $\mathrm{Hom}_R(M, X) = 0$. A similar result holds for the module $(0, Y)$.

Proposition 3.4.1 can be reformulated as follows. Let H be the functor defined in Sect. 3.1. We will see soon that each injective K-module is isomorphic to the module $H(X, Y)$ for some injective R-module X and some injective S-module Y.

Let (A, B) be a K-module and f', g' the homomorphisms defined at the beginning of Sect. 3.1. These homomorphisms play a special role. They correspond to the homomorphisms f, g of module multiplication for the isomorphisms mentioned at the beginning of Sect. 3.1. It is more convenient to write f and g instead of f' and g', respectively. Thus, f is an S-homomorphism $B \to \mathrm{Hom}_R(M, A)$, where $f(b)(m) = mb$, $b \in B$, $m \in M$, and g is an R-homomorphism $A \to \mathrm{Hom}_S(N, B)$, where $g(a)(n) = na$, $a \in A$, $n \in N$. We have an exact sequence

$$0 \to L(A) \to A \xrightarrow{g} \mathrm{Hom}_S(N, B),$$

$$0 \to L(B) \to B \xrightarrow{f} \mathrm{Hom}_R(M, A)$$

of R-modules and S-modules, respectively, where $L(A)$, $L(B)$ are the submodules defined in Sect. 3.2. The mappings

$$(1, f) \colon (A, B) \to (A, H(A)), \quad (g, 1) \colon (A, B) \to (H(B), B)$$

are K-homomorphisms (see Lemma 3.1.3 and the remarks after Corollary 3.1.4); they will be often used. In addition, if X is an R-module, $(X, H(X))$ is a K-module and I is the trace ideal of the ring K, then $L(X) = \{x \in X \mid Ix = 0\}$, $L(H(X)) = 0$.

We pass to the description of injective K-modules. Here we consider two informative cases. Later, we will show that the general case can be reduced to these two cases.

Theorem 3.4.2 *Let (A, B) be a module with $L(A) = 0 = L(B)$. The module (A, B) is injective if and only if A and B are injective modules. In addition, f and g are isomorphisms.*

Proof Let (A, B) be an injective module. It follows from the assumption that f and g are monomorphisms. Therefore, $(1, f)$ is a monomorphism. Since the module (A, B) is injective, the image $\mathrm{Im}(1, f)$ is a direct summand in $(A, H(A))$. The complement summand has the form $(0, Z)$, where Z is some direct summand in $H(A)$. Consequently, $MZ = 0$ and $Z \subseteq L(H(A))$. As noted above, $L(H(A)) = 0$ and $Z = 0$. Thus, $(1, f)$ is an isomorphism. Therefore, $(A, H(A))$ is an injective module and f is isomorphism. By Proposition 3.4.1, the module A is injective. Similarly, it is proved that g is an isomorphism and the module B is injective.

We assume now that A and B are injective modules. Again, we use the homomorphisms $(1, f)$ and $(g, 1)$. We consider the K-module $(\mathrm{Hom}_S(N, H(A)), H(A))$ and the K-homomorphism

$$(h, 1) \colon (A, H(A)) \to (\mathrm{Hom}_S(N, H(A)), H(A))$$

described in Lemma 3.1.3. Via direct calculations, one can verify that $h = g f_*$, where $f_* \colon \mathrm{Hom}_S(N, B) \to \mathrm{Hom}_S(N, H(A))$ is a homomorphism induced by the homomorphism f. Since f_* and g are monomorphisms, h is a monomorphism. By Proposition 3.4.1, the module $(A, H(A))$ is injective. By repeating arguments from the first part the proof, we can verify that h is an isomorphism. Consequently, the monomorphism f_* is an epimorphism. Therefore, f_* and g are isomorphisms.

Similarly, we can prove that f is an isomorphism. Thus, $(1, f): (A, B) \to (A, H(A))$ is an isomorphism and the module (A, B) is injective, which is what we required. □

In any module (A, B), the subsets $(L(A), 0)$, $(0, L(B))$ and $(L(A), L(B))$ are submodules. In the setting of injective K-modules, one more important situation occurs: where $(L(A), L(B))$ is an essential submodule of (A, B). (This the case if the trace ideals of the ring K are equal to the zero ideal.)

We recall several notions from ring theory. Let V be a module over some ring and G, Z two submodules of V. The submodule G is said to be *closed* (in V) if G does not have proper essential extensions in V. A submodule G is called a *closure* of the submodule Z (in V) if $Z \subseteq G$, G is an essential extension of the module Z and G is closed in V. In V, every submodule has at least one closure, which is not necessarily unique. The symbol $\overline{Z}$ denotes some closure of the submodule Z.

Theorem 3.4.3 *Let (A, B) be a module such that $(L(A), L(B))$ is an essential submodule of (A, B). The module (A, B) is injective if and only if there exist closures $\overline{L(A)}$ and $\overline{L(B)}$ such that $\overline{L(A)}$ is an injective R-module and $\overline{L(B)}$ is an injective S-module,*

$$L(A) \cap M\overline{L(B)} = 0, \qquad N\overline{L(A)} \cap L(B) = 0,$$

$$\mathrm{Hom}_R(M, \overline{L(A)}) \subseteq \mathrm{Im}\, f, \qquad \mathrm{Hom}_S(N, \overline{L(B)}) \subseteq \mathrm{Im}\, g.$$

Proof We assume that the module (A, B) is injective. There exist a closure (A_1, B_2) of the submodule $(L(A), 0)$ and a closure (A_2, B_1) of the submodule $(0, L(B))$ such that $(A_1, B_2) \cap (A_2, B_1) = 0$. Any closed submodule of an injective module is injective. Consequently, $(A_1, B_2) \oplus (A_2, B_1)$ is an essential injective submodule of (A, B); therefore, $(A, B) = (A_1, B_2) \oplus (A_2, B_1)$. One can directly verify that A_1 is an essential extension of the module $L(A)$ and B_1 is an essential extension of the module $L(B)$. The direct summands A_1 and B_1 are closed submodules. Therefore, $A_1 = \overline{L(A)}$ and $B_1 = \overline{L(B)}$. By Lemma 3.1.3, we have the homomorphism $(1, f): (A_1, B_2) \to (A_1, H(A_1))$ (more precisely, we have to take the restriction of f to B_2 instead of f). Since $B_2 \cap L(B) = 0$, we have that $(1, f)$ is a monomorphism. Since the module (A_1, B_2) is injective, we can repeat the argument from the proof of Theorem 3.4.2 and obtain that $(1, f)$ is an isomorphism. Consequently, the module $(A_1, H(A_1))$ is injective. Then the module A_1 is injective by Proposition 3.4.1. In addition, $\mathrm{Hom}_R(M, A_1) \subseteq \mathrm{Im}\, f$. Thus, it follows from the inclusions $L(A) \subseteq A_1$ and $MB_1 \subseteq A_2$ that $L(A_1) \cap MB_1 = 0$. The remaining assertions are similarly proved; in particular, $(g, (1): (A_2, B_1) \to (H(B_1), B_1)$ is an isomorphism.

Now we assume that there exist closures $\overline{L(A)}$ and $\overline{L(B)}$ such that they are injective, $L(A) \cap M\overline{L(B)} = 0$, $N\overline{L(A)} \cap L(B) = 0$, $\mathrm{Hom}_R(M, \overline{L(A)}) \subseteq \mathrm{Im}\, f$ and $\mathrm{Hom}_S(N, \overline{L(B)}) \subseteq \mathrm{Im}\, g$. Set $A_1 = \overline{L(A)}$ and $B_1 = \overline{L(B)}$. We also consider submodules (A_1, NA_1) and (MB_1, B_1). We remark that their intersection is equal to the zero module. We also take $(A_1, H(A_1))$ and $(H(B_1), B_1)$, which are injective modules by Proposition 3.4.1. We also consider the homomorphisms

$$(1, f): (A_1, NA_1) \to (A_1, H(A_1)), \quad (g, 1): (MB_1, B_1) \to (H(B_1), B_1),$$

where f and g denote the restrictions of the homomorphisms to the corresponding submodules. In fact, they are monomorphisms, since $\operatorname{Ker} f = L(B) \cap NA_1 = 0$; similar relations hold for $\operatorname{Ker} g$. The sum of mappings $(1, f) + (g, 1)$ can be extended to a monomorphism $(A, B) \to (A_1, H(A_1)) \oplus (H(B_1), B_1)$. We identify (A, B) with the image of this monomorphism. We note that the role of the submodule $L(A)$ in the above sum is similar to its role in (A, B).

We have direct decompositions

$$A = A_1 \oplus A_2, \quad B = B_1 \oplus B_2, \quad \text{where} \quad A_2 = A \cap H(B_1), \quad B_2 = B \cap H(A_1).$$

It is clear that there exist K-modules (A_1, B_2), (A_2, B_1) and a direct decomposition $(A, B) = (A_1, B_2) \oplus (A_2, B_1)$. It is convenient to return to the original module (A, B) and to assume that the given decomposition is a decomposition of this module. We again take the monomorphism $(1, f): (A_1, B_2) \to (A_1, H(A_1))$. Let $\alpha \in H(A_1)$. Since $H(A_1) \subseteq \operatorname{Im} f$, we have that $\alpha = f(b)$ for some $b \in B$. Then $b = c + d$, where $c \in B_2, d \in B_1$. For every $m \in M$, we have

$$\alpha(m) = f(b)(m) = mb = mc + md, \quad mc, mb \in A_1, \quad md \in A_2.$$

Therefore, $md = 0$ and $mb = mc$. Then we obtain

$$\alpha(m) = mc = f(c)(m), \quad \alpha = f(c).$$

We have proved that $(1, f)$ is an isomorphism. Thus, $(A, B) \cong (A_1, H(A_1)) \oplus (H(B_1), B_1)$ and the module (A, B) is injective. $\qquad \square$

Corollary 3.4.4 *Under the conditions and the notation of Theorem 3.4.3, we have the canonical isomorphism*

$$(A, B) \cong (A_1, H(A_1)) \oplus (H(B_1), B_1).$$

Remark Let K be an arbitrary formal matrix ring and (A, B) a K-module. Then (A, B) is an essential extension of the module $(L(A) + MB, L(B) + NA)$. In addition, (A, B) is an essential extension of the module $(L(A), L(B))$ provided K is a ring with zero trace ideals. Indeed, let $(a, b) \in (A, B)$ and $a \neq 0$. If $na = 0$ for all $n \in N$, then $a \in L(A)$; if $na \neq 0$ for some $n \in N$, then $\begin{pmatrix} 0 & 0 \\ n & 0 \end{pmatrix} (a, b) = (0, na) \in (0, NA)$. The case $b \neq 0$ is considered similarly. If K is a ring with zero trace ideals, then

$$MB \subseteq L(A), \quad NA \subseteq L(B),$$
$$L(A) + MB = L(A), \quad L(B) + NA = L(B).$$

Corollary 3.4.5 *Let (A, B) be a K-module.*

(1) *If K is a ring with zero trace ideals, then (A, B) is an injective module if and only if $L(A)$ and $L(B)$ are injective modules, $\operatorname{Hom}_R(M, L(A)) \subseteq \operatorname{Im} f$ and $\operatorname{Hom}_S(N, L(B)) \subseteq \operatorname{Im} g$.*

(2) *If $K = \begin{pmatrix} R & M \\ 0 & S \end{pmatrix}$ is a ring of triangular matrices, then (A, B) is an injective module if and only if A and $L(B)$ are injective modules and $f: B \to Hom_R(M, A)$ is an epimorphism.*

Proof (1). Since K is a ring with zero trace ideals, the relations $L(X) = X$, $L(H(X)) = 0$ hold for any module $(X, H(X))$. Similar relations hold for the modules of the form $(H(Y), Y)$.

We assume that the module (A, B) is injective. We identify it with the isomorphic image from Corollary 3.4.4. Then $L(A) = \overline{L(A)}$ and $L(B) = \overline{L(B)}$. Therefore, $L(A)$ and $L(B)$ are injective modules. The remaining part (1) follows from Theorem 3.4.3.

(2). The assertion follows from (1) and the relation $L(A) = A$. □

Theorem 3.4.6 *An arbitrary K-module (A, B) is injective if and only if some closure of the submodule $(L(A), L(B))$ is injective and there exist closures $\overline{L(A)}$, $\overline{L(B)}$ such that the factor modules $A / \left(\overline{L(A)} + g^{-1} H \left(\overline{L(B)} \right) \right)$ and $B / \left(\overline{L(B)} + f^{-1} H \left(\overline{L(A)} \right) \right)$ are injective.*

Proof We assume that the module (A, B) is injective. Any closed submodules of an injective module is injective. There exists a direct decomposition $(A, B) = (G, H) \oplus (C, D)$, where the first summand is some closure of the submodule $(L(A), L(B))$. We can apply Theorem 3.4.3 to the module (G, H). Consequently, there exist injective closures A_1, B_1 of the modules $L(A), L(B)$, respectively. There exists a direct decomposition $(G, H) = (A_1, B_2) \oplus (A_2, B_1)$ for some submodules A_2, B_2. In addition, it follows from the proof of Theorem 3.4.3 that the mappings

$$(1, f): (A_1, B_2) \to (A_1, H(A_1)), \qquad (g, 1): (A_2, B_1) \to (H(B_1), B_1)$$

are isomorphisms. The module (C, D) is injective and $L(C) = 0 = L(D)$. By Theorem 3.4.2, C and D are injective modules. It remains to note that

$$C \cong A/(A_1 \oplus A_2) = A/ \left(A_1 + g^{-1} H(B_1) \right).$$

Similar relations hold for other factor modules.

We assume that the conditions of the theorem hold. Then $(A, B) = (G, H) \oplus (C, D)$ for some closure (G, H) of the submodule $(L(A), L(B))$ and a module (C, D) such that $L(C) = 0 = L(D)$. We can apply Theorem 3.4.3 to the module (G, H). As above, there exists a direct decomposition $(G, H) = (A_1, B_2) \oplus (A_2, B_1)$, and mappings $(1, f)$, $(g, 1)$ are isomorphisms. By repeating the above argument, we obtain that C and D are injective modules. By Theorem 3.4.2, the module (C, D) is injective. Therefore, the module (A, B) is injective. $\qquad\square$

From the above theorems we conclude the following.

Remark Every injective module (A, B) has the direct decomposition

$$(A, B) = (A_1, B_2) \oplus (A_2, B_1) \oplus (C, D),$$
$$A_1 = \overline{L(A)}, \quad B_1 = \overline{L(B)}, \quad L(C) = 0 = L(D),$$

and the canonical mappings

$$B_2 \to \operatorname{Hom}_R(M, A_1), \quad A_2 \to \operatorname{Hom}_S(N, B_1),$$
$$D \to \operatorname{Hom}_R(M, C), \quad C \to \operatorname{Hom}_S(N, D)$$

are isomorphisms.

Corollary 3.4.7 *A module (A, B) is injective if and only if there exist an injective R-module X and an injective S-module Y such that $(A, B) \cong (X, H(X)) \oplus (Y, H(Y))$.*

Corollary 3.4.7 follows from the previous remark, Theorems 3.4.2 and 3.4.3. $\quad\square$

From the obtained information about injective modules, we can obtain a description of injective hulls. The injective hull of some module V is denoted by $\widehat{V}$.

Lemma 3.4.8 *Let V be a module over some ring and C_1, C_2 two closed submodules of V such that $C_1 \cap C_2 = 0$ and $C_1 \oplus C_2$ is an essential submodule of V. Then*

$$\widehat{V} = \widehat{C_1} \oplus \widehat{C_2}, \quad \text{where} \quad \widehat{C_1} \cong \widehat{V/C_2}, \quad \widehat{C_2} \cong \widehat{V/C_1}.$$

Proof We have $\widehat{V} \cong \widehat{C_1} \oplus \widehat{C_2}$. Then $C_2 \cong (C_1 \oplus C_2)/C_1 \subseteq V/C_1$ and V/C_1 is an essential extension of the module $(C_1 \oplus C_2)/C_1$. Indeed, let B be a submodule such that $C_1 \subseteq B$ and $C_1 \neq B$. Since C_1 is a closed submodule of V, we have $B \cap C_2 \neq 0$. Consequently, $B/C_1 \cap (C_1 \oplus C_2)/C_1 \neq 0$. Therefore, the module C_2 is isomorphic to an essential submodule of V/C_1, whence $\widehat{C_2} \cong \widehat{V/C_1}$. The second isomorphism can be similarly proved. $\qquad\square$

Corollary 3.4.9 *Let (A, B) be a K-module.*

(1) *If $L(A) = 0 = L(B)$, then there exists a K-module $(\widehat{A}, \widehat{B})$, and this module is the injective hull of the module (A, B). In addition, we have canonical isomorphisms $\widehat{A} \cong \operatorname{Hom}_S(N, \widehat{B})$ and $\widehat{B} \cong \operatorname{Hom}_R(M, \widehat{A})$.*

(2) *If (A, B) is an essential extension of the module $(L(A), L(B))$, then the module $\left(\widehat{L(A)}, H\left(\widehat{L(A)}\right)\right) \oplus \left(H\left(\widehat{L(B)}\right), \widehat{L(B)}\right)$ is the injective hull of the module (A, B).*

(3) *The injective hull of the module* (A, B) *has the form* $U \oplus V$, *where* U *is the injective hull of the module* $(L(A), L(B))$, *and there exists a closure* W *of the submodule* $(L(A), L(B))$ *such that* V *is the injective hull of the factor module* $(A, B)/W$. *The module* $(L(A), L(B))$ *satisfies the conditions of Item (2) and the module* $(A, B)/W$ *satisfies the conditions of Item (1).*

Proof (1). We consider the K-module $(\widehat{A}, H(\widehat{A}))$. By Proposition 3.4.1, this module is injective. Since $H(A) \subseteq H(\widehat{A})$, we can consider the homomorphism $(1, f)$: $(A, B) \to (\widehat{A}, H(\widehat{A}))$. Since $L(B) = 0$, we have that $(1, f)$ is a monomorphism. Its image is an essential submodule of $(\widehat{A}, H(\widehat{A}))$. Otherwise, $(\widehat{A}, H(\widehat{A}))$ contains some injective hull of the image which is of the form $(\widehat{A}, Y)$ for some proper submodule Y. Then $H(\widehat{A}) = Y \oplus Z$, where $Z \neq 0$ and $MZ = 0$. This contradicts the relation $L(H(\widehat{A})) = 0$; this argument is similar to the argument from the beginning of the proof of Theorem 3.4.2. Thus, $(\widehat{A}, H(\widehat{A}))$ is the injective hull of the module (A, B). Similarly, the module $(H(\widehat{B}), \widehat{B})$ is also the injective hull of the module (A, B). We identify the module (A, B) with its isomorphic image in these two hulls. Then the identity mapping of the module (A, B) can be extended to an isomorphism (α, β): $(\widehat{A}, H(\widehat{A})) \to (H(\widehat{B}), \widehat{B})$. In addition, α is an extension of the monomorphism $g \colon A \to \mathrm{Hom}_S(N, B)$, and β is an extension of the converse isomorphism to $f \colon B \to \mathrm{Im}\, f$. This is what is meant by the canonical nature of the isomorphisms from the corollary.

(2). The injective hull of the module (A, B) coincides with the injective hull of $(L(A), L(B))$, which is equal to $(L(A), 0) \oplus (0, L(B))$. It follows from the proof of Theorem 3.4.3 that the injective hull of the module $(L(A), 0) \oplus (0, L(B))$ coincides with the sum from the corollary.

(3). The assertion is verified with the use of Lemma 3.4.8.   □

3.5 Maximal Rings of Fractions

We apply the results and methods of the previous section to the description of maximal rings of fractions of formal matrix rings of order 2. In some cases, we will find their precise structure.

As above, the injective hull of an arbitrary module V is denoted by $\hat{V}$. If C, D are submodules of a T-module A such that $C \cap D = 0$ and $C + D$ is an essential submodule of A, then D is called a *complement* to C (in A). Set $L = \mathrm{End}_T A$. Then A is a T-L-bimodule, and the ring $\mathrm{End}_L A$ is called the *biendomorphism ring* of the module A.

The maximal (left) *ring of fractions* of the ring T is defined as the biendomorphism ring of the injective hull of the module $_T T$. It is denoted by $Q(T)$.

We return to Corollary 3.4.9 concerning the structure of the injective hull of a K-module (A, B). Firstly, under the conditions of Corollary 3.4.9(1), there are canonical isomorphisms

$$(\hat{A}, H(\hat{A})) \cong (\hat{A}, \hat{B}) \cong (H(\hat{B}), \hat{B}),$$

and the condition of Corollary 3.4.9(2) holds provided the trace ideals I and J of the ring K are nilpotent.

Indeed, let $(a, b) \in (A, B)$ and $(a, b) \notin (L(A), L(B))$. We assume that $a \notin L(A)$; we can also assume that $b = 0$. There exists an element $n_1 \in N$ with $n_1 a \neq 0$. If $n_1 a \notin L(B)$, then $(m_1 n_1)a \neq 0$ for some $m_1 \in M$, where $m_1 n_1 \in I$. We can use the same argument for the element $(m_1 n_1)a$. Since the ideal I is nilpotent, this process will stop after finitely many steps. The case $b \notin L(B)$ is similar.

In Corollary 3.4.9(3), the module V is not uniquely constructed. We describe a slightly different construction. Unfortunately, this approach does not provide a canonical method of choosing the summand V.

Let (C, D) be some complement to the submodule $(L(A), L(B))$ in (A, B). In Corollary 3.4.9(3), we can take the injective hull of the module (C, D) as the module V. Since $L(C) = 0 = L(D)$, it follows from Corollary 3.4.9(1) that this hull is isomorphic to $(\hat{C}, \hat{D})$, $\hat{C} \cong H(\hat{D})$, $\hat{D} \cong H(\hat{C})$. Thus, for the injective hull of the module (A, B), there exists an isomorphism

$$\widehat{(A, B)} \cong (\widehat{L(A)}, H(\widehat{L(A)})) \oplus (H(\widehat{L(B)}), \widehat{L(B)}) \oplus (\hat{C}, \hat{D}).$$

We also have isomorphisms

$$\widehat{(A, B)} \cong (\widehat{L(A) \oplus C}, H(\widehat{L(A) \oplus C})) \oplus (H(\widehat{L(B)}), \widehat{L(B)})$$
$$\cong (\widehat{L(A)}, H(\widehat{L(A)})) \oplus (H(\widehat{L(B) \oplus D}), \widehat{L(B) \oplus D}).$$

There are informative special cases where the information about the complement (C, D) is not necessary to find the injective hull $\widehat{(A, B)}$.

We assume that $L(A) = 0$. Let D be a complement to $L(B)$. It is not difficult to verify that (MD, D) is a complement to $(0, L(B))$ in (A, B). Consequently, there exists an isomorphism $\widehat{(A, B)} \cong (H(\widehat{B}), \widehat{B})$. If $L(B) = 0$, then $\widehat{(A, B)} \cong (\hat{A}, H(\hat{A}))$.

Thus, we can always represent the injective hull E of the module (A, B) in the form $(X, H(X)) \oplus (H(Y), Y)$, where X is some injective R-module and Y is some injective S-module; see also Corollary 3.4.7.

We find the endomorphism ring and the biendomorphism ring of a K-module of the form $(X, H(X)) \oplus (H(Y), Y)$ for arbitrary modules X and Y. From Lemma 3.1.3, we obtain the following lemma.

Lemma 3.5.1 *The endomorphism ring of the K-module $E = (X, H(X)) \oplus (H(Y), Y)$ is isomorphic to the formal matrix ring*

$$T = \begin{pmatrix} \mathrm{End}_R X & \mathrm{Hom}_S(H(X), Y) \\ \mathrm{Hom}_R(H(Y), X) & \mathrm{End}_S Y \end{pmatrix}.$$

As a right T-module, E is a module of row vectors $(X \oplus H(X), H(Y) \oplus Y)$. This module can also be represented in the form of the direct sum

$$E = (X, H(Y)) \oplus (H(X), Y) = G \oplus F.$$

Therefore, we obtain the following result.

Lemma 3.5.2 *The biendomorphism ring of the K-module $E = (X, H(X)) \oplus (H(Y), Y)$ is isomorphic to the formal matrix ring*

$$\begin{pmatrix} \mathrm{End}_T\, G & \mathrm{Hom}_T(F, G) \\ \mathrm{Hom}_T(G, F) & \mathrm{End}_T\, F \end{pmatrix}.$$

We focus on the special cases, where $X = 0$ or $Y = 0$. These cases arise if we take the injective hull of the module (A, B) under the condition $L(A) = 0$ or $L(B) = 0$.

Thus, let $E = (X, H(X))$. Then $\mathrm{End}_K E \cong \mathrm{End}_R X = T$. The right T-module $(X, H(X))$ can be represented as $X \oplus H(X)$. Consequently, the biendomorphism ring $\mathrm{End}_A E$ is the formal matrix ring

$$\begin{pmatrix} \mathrm{End}_T\, X & \mathrm{Hom}_T(H(X), X) \\ \mathrm{Hom}_T(X, H(X)) & \mathrm{End}_T\, H(X) \end{pmatrix}.$$

The biendomorphism ring of the K-module $(H(Y), Y)$ has a similar structure.

Now we take the left K-module $_K K = \begin{pmatrix} (R, M) \\ (N, S) \end{pmatrix}$ as the module (A, B); see Sect. 3.1. The submodule $L(R, M)$ is equal to $(\mathrm{Ann}_R N, \mathrm{Ann}_M N)$, where $\mathrm{Ann}_R N = \{r \in R \mid Nr = 0\}$ is the annihilator of the R-module N, $\mathrm{Ann}_M N = \{m \in M \mid Nm = 0\}$. We denote the submodule $L(R, M)$ by L_1. The submodule $L(N, S)$ has a similar structure; we denote it by L_2. Further, let (C, D) be some complement to the submodule (L_1, L_2) in $_K K$. The injective hull E of the module $_K K$ has the form

$$(\widehat{L_1 \oplus \widehat{C}}, H(\widehat{L_1 \oplus \widehat{C}})) \oplus (H(\widehat{L_2}), \widehat{L_2})$$

or

$$(\widehat{L_1}, H(\widehat{L_1})) \oplus (H(\widehat{L_2 \oplus \widehat{D}}), \widehat{L_2 \oplus \widehat{D}}).$$

Let $T = \mathrm{End}_K E$ be the endomorphism ring of the module E in the first or the second form. Then we can represent the right T-module E in the form

$$(\widehat{L_1 \oplus \widehat{C}}, H(\widehat{L_2})) \oplus (H(\widehat{L_1 \oplus \widehat{C}}), \widehat{L_2}) = G \oplus F$$

or

$$(\widehat{L_1}, H(\widehat{L_2 \oplus \widehat{D}})) \oplus (H(\widehat{L_1}), \widehat{L_2 \oplus \widehat{D}}) = G' \oplus F'.$$

Corollary 3.5.3 *Let K be a ring and Q the maximal ring of fractions of K, i.e., the ring $\mathrm{End}_T E$. Then either Q is isomorphic to the formal matrix ring*
$$\begin{pmatrix} \mathrm{End}_T G & \mathrm{Hom}_T(F, G) \\ \mathrm{Hom}_T(G, F) & \mathrm{End}_T F \end{pmatrix} \ \textit{or } Q \textit{ is isomorphic to a similar ring if we replace}$$
G and F by G' and F', respectively.

In a somewhat different form, the ring $Q(K)$ is found in [91].

Similar to the case of an arbitrary module (A, B), a problem remains; it is related to the non-canonical choice of the complement (C, D). We remarked above that if $L(A) = 0$ or $L(B) = 0$, then the submodule (C, D) is not used explicitly for the construction of the injective hull of the module (A, B). We apply this fact to the K-module K. We draw attention to the fact that this module can be represented in the form of the direct sum $(R, N) \oplus (M, S)$. Therefore, the following four cases are possible.

(1) N_R and M_S are exact modules.

Then $L(R) = 0$ and $L(S) = 0$ in the modules (R, N) and (M, S), respectively.

(2) $\mathrm{Ann}_M N = 0 = \mathrm{Ann}_N M$.

Then $L(N) = 0$ and $L(M) = 0$ in the modules (R, N) and (M, S), respectively.

(3) $L(R) = 0$ and $L(M) = 0$.

(4) $L(N) = 0$ and $L(S) = 0$.

In case (1), it follows from the previous material that

$$(\widehat{R, N}) \cong (H(\widehat{N}), \widehat{N}), \quad (\widehat{M, S}) \cong (\widehat{M}, H(\widehat{M})).$$

Then we obtain

$$_K\widehat{K} \cong (\widehat{M}, H(\widehat{M})) \oplus (H(\widehat{N}), \widehat{N}) = E.$$

Let $T = \mathrm{End}_K E$. Then the right T-module E has the form

$$(\widehat{M}, H(\widehat{N})) \oplus (H(\widehat{M}), \widehat{N}) = G \oplus F.$$

The maximal ring of fractions $Q(K)$ (i.e., $\mathrm{End}_T E$) has the form specified in Corollary 3.5.3.

In the case (2), we have

$$_K\widehat{K} \cong (\widehat{R}, H(\widehat{R})) \oplus (H(\widehat{S}), \widehat{S}) = E, \quad T = \mathrm{End}_K E,$$
$$E_T = (\widehat{R}, H(\widehat{S})) \oplus (H(\widehat{R}), \widehat{S}) = G \oplus F,$$

and, finally, $Q(K) = \mathrm{End}_T E$.

In cases (3) and (4), the ring $Q(K)$ has a similar description; see [99].

The following question is of interest. When can we put $Q(R)$ and $Q(S)$ on the main diagonal of the ring $Q(K)$ from Corollary 3.5.3? We consider one situation where this question allows a satisfactory answer. Namely, we assume that N_R and M_S are faithful modules; i.e., $L(R) = 0 = L(S)$ and $L(M) = 0 = L(N)$. Then $L(R, M) = 0 = L(N, S)$ and the injective hull E of the module $_K K$ can be represented in the form $((\hat{R}, \hat{M}), (\hat{N}, \hat{S}))$ or in the form of the direct sum $(\hat{R}, \hat{N}) \oplus (\hat{M}, \hat{S})$. Then there are canonical isomorphisms

$$H(\hat{N}) \cong \hat{R}, \quad H(\hat{R}) \cong \hat{N},$$
$$H(\hat{S}) \cong \hat{M}, \quad H(\hat{M}) \cong \hat{S}.$$

From these isomorphisms and Lemma 3.1.3, we obtain the following isomorphisms for endomorphism rings and homomorphism groups.

$$\operatorname{End}_R \hat{R} \cong \operatorname{End}_K (\hat{R}, \hat{N}) \cong \operatorname{End}_S \hat{N},$$
$$\operatorname{End}_R \hat{M} \cong \operatorname{End}_K (\hat{M}, \hat{S}) \cong \operatorname{End}_S \hat{S},$$
$$\operatorname{Hom}_R(\hat{R}, \hat{M}) \cong \operatorname{Hom}_K ((\hat{R}, \hat{N}), (\hat{M}, \hat{S})) \cong \operatorname{Hom}_S(\hat{N}, \hat{S}),$$
$$\operatorname{Hom}_R(\hat{M}, \hat{R}) \cong \operatorname{Hom}_K ((\hat{M}, \hat{S}), (\hat{R}, \hat{N})) \cong \operatorname{Hom}_S(\hat{S}, \hat{N}).$$

If necessary, we will identify the corresponding isomorphic rings and modules.

The endomorphism ring of the K-module $E = ((\hat{R}, \hat{M}), (\hat{N}, \hat{S}))$ is isomorphic to the ring $\operatorname{End}_R(\hat{R} \oplus \hat{M})$, i.e., it is isomorphic to the formal matrix ring

$$\begin{pmatrix} \operatorname{End}_R \hat{R} & \operatorname{Hom}_R(\hat{R}, \hat{M}) \\ \operatorname{Hom}_R(\hat{M}, \hat{R}) & \operatorname{End}_R \hat{M} \end{pmatrix}. \tag{$*$}$$

This ring is also isomorphic to the ring $\operatorname{End}_S(\hat{N} \oplus \hat{S})$, i.e., it is isomorphic to the formal matrix ring

$$\begin{pmatrix} \operatorname{End}_S \hat{N} & \operatorname{Hom}_S(\hat{N}, \hat{S}) \\ \operatorname{Hom}_S(\hat{S}, \hat{N}) & \operatorname{End}_S \hat{S} \end{pmatrix}. \tag{$**$}$$

We denote by $\begin{pmatrix} A & V \\ W & B \end{pmatrix}$ the endomorphism ring $\operatorname{End}_K E$ represented in the form $(*)$ or $(**)$.

We need some standard information related to dualities.

Let R, S, T be arbitrary rings and U an R-S-bimodule. Further, let

$$H' = \operatorname{Hom}_R(\_, U): R\text{-mod} \to \operatorname{Mod}\text{-}S,$$
$$H'' = \operatorname{Hom}_S(\_, U): \operatorname{Mod}\text{-}S \to R\text{-mod}$$

be contravariant functors. For every left R-module X, there exists a natural homomorphism

$$\sigma_X : X \to H''H'(X), \quad \sigma_X(x)(f) = f(x),$$
$$x \in X, \quad f \in H'(X)$$

which is called the *evaluation map*.

We denote by H each contravariant functor of the form H' or H''. If X is an R-T-bimodule, then σ_X is a T-module homomorphism. If σ_X is an isomorphism, then the module X is said to be U-*reflexive*.

Now we give a partial answer to the above question: when do there exist canonical isomorphisms $\mathrm{End}_T\, G \cong Q(R)$ and $\mathrm{End}_T\, H \cong Q(S)$? In addition, is it possible to calculate the groups standing on the secondary diagonal in the matrix ring from Corollary 3.5.3?

We return to the right T-modules $(\hat{R}, \hat{M}) \oplus (\hat{N}, \hat{S})$. There exists the right T-module $(\hat{R}, H(\hat{R}))$ (we mean the right analogue of the construction of the K-module $(X, H(X))$ from Lemma 3.1.3; see the remark 2 in Sect. 3.1). Here $H(\hat{R}) = \mathrm{Hom}_A(W, \hat{R})$, $W = \mathrm{Hom}_R(\hat{M}, \hat{R})$ and $A = \mathrm{End}_R\, \hat{R}$. By the right-side analogue of Lemma 3.1.3, there exists a canonical T-module homomorphism

$$(1, \gamma) : (\hat{R}, \hat{M}) \to (\hat{R}, H(\hat{R})),$$

where $\gamma(x)(\alpha) = \alpha(x)$ for $x \in \hat{M}$, $\alpha \in \mathrm{Hom}_R(\hat{M}, \hat{R})$. Thus, γ coincides with the evaluation map

$$\sigma_{\hat{M}} : \hat{M} \to \mathrm{Hom}_A(\mathrm{Hom}_R(\hat{M}, \hat{R}), \hat{R}) = H^2(\hat{M}).$$

Consequently, if $_R\hat{M}$ is an $\hat{R}$-reflexive module, then γ and $(1, \gamma)$ are isomorphisms and the T-modules $(\hat{R}, \hat{M})$, $(\hat{R}, H(\hat{R}))$ are isomorphic. Then it follows from the right-side analogue of Corollary 3.1.4 that

$$\mathrm{End}_T(\hat{R}, \hat{M}) \cong \mathrm{End}_A\, \hat{R} = Q(R).$$

We can apply a similar argument to T-modules $(\hat{N}, \hat{S})$. If $\hat{N}_S$ is an $\hat{S}$-reflexive module, then $(\hat{N}, \hat{S}) \cong (H(\hat{S}), \hat{S})$ (the right analogue of the construction $(H(Y), Y)$ from Lemma 3.1.3) and

$$\mathrm{End}_T(\hat{N}, \hat{S}) \cong \mathrm{End}_B\, \hat{S} = Q(S).$$

By the right-side analogue of Lemma 3.1.3, there are also isomorphisms

$$\mathrm{Hom}_T(G, F) \cong \mathrm{Hom}_B(\hat{M}, \hat{S}) \cong \mathrm{Hom}_B(H(\hat{S}), \hat{S})$$
$$\cong \mathrm{Hom}_B(\mathrm{Hom}_S(N, \hat{S}), \hat{S}) = H^2(N).$$

Similarly, $\mathrm{Hom}_T(F, G) \cong H^2(M)$. Thus, we obtain the following corollary.

Corollary 3.5.4 *If $\hat{M}$ is an $\hat{R}$-reflexive module and $\hat{N}$ is an $\hat{S}$-reflexive module, then there exists an isomorphism*

$$Q(K) \cong \begin{pmatrix} Q(R) & H^2(M) \\ H^2(N) & Q(S) \end{pmatrix}.$$

If

$$\text{Hom}_R(\hat{M}/M, \hat{R}) = 0 = \text{Hom}_S(\hat{N}/N, \hat{S}),$$

then it is not difficult to verify that $H^2(M) \cong \hat{M}$ and $H^2(N) \cong \hat{N}$.

The R-module $\hat{M}$ is $\hat{R}$-reflexive if $\hat{M}$ is a *finitely $\hat{R}$-projective* module, i.e., $\hat{M}$ is a direct summand of a finite direct sum of copies of the module $\hat{R}$; see Sect. 5.2 for more about such modules. If M is a finitely generated projective R-module then the last property holds. A similar assertion holds for the S-module $\hat{N}$.

Again, we consider a T-module $(\hat{R}, \hat{M})$. There also exists a right T-module $(H(\hat{M}), \hat{M})$. By the right-side analogue of Lemma 3.1.3, there exists a canonical T-module homomorphism

$$(\delta, (1)): (\hat{R}, \hat{M}) \to (H(\hat{M}), \hat{M}),$$

and δ coincides with the evaluation map

$$\sigma_{\hat{R}}: \hat{R} \to \text{Hom}_B(\text{Hom}_R(\hat{R}, \hat{M}), \hat{M}) = H^2(\hat{R}).$$

If the R-module $\hat{R}$ is $\hat{M}$-reflexive, then δ, $(\delta, 1)$ are isomorphisms and End_T $(\hat{R}, \hat{M}) \cong \text{End}_B \hat{M}$, where the last ring is the biendomorphism ring of the R-module $\hat{M}$. Similarly, if the S-module $\hat{S}$ is $\hat{N}$-reflexive, then $(\hat{N}, \hat{S}) \cong (\hat{N}, H(\hat{N}))$ and $\text{End}_T(\hat{N}, \hat{S}) \cong \text{End}_A \hat{N}$, where $\text{End}_A \hat{N}$ is the biendomorphism ring of the S-module $\hat{N}$.

Let $\hat{R}$ be an $\hat{M}$-reflexive module and $\hat{S}$ an $\hat{N}$-reflexive module. By the right-side analogue of Lemma 3.1.3, there are isomorphisms

$$\text{Hom}_T(G, F) \cong \text{Hom}_A(\hat{R}, \hat{N}) \cong \text{Hom}_A(H(\hat{N}), \hat{N}) = H^2(N).$$

Similarly, we have $\text{Hom}_T(F, G) \cong H^2(M)$. As a result, we obtain the following result.

Corollary 3.5.5 *If $\hat{R}$ is an $\hat{M}$-reflexive module and $\hat{S}$ is an $\hat{N}$-reflexive module, then*

$$Q(K) \cong \begin{pmatrix} \text{End}_B \hat{M} & H^2(M) \\ H^2(N) & \text{End}_A \hat{N} \end{pmatrix}.$$

The R-module $\hat{R}$ is $\hat{M}$-reflexive if $\hat{R}$ is a finitely $\hat{M}$-projective module; in particular, M is a generator R-module (generator modules are defined in Sect. 3.8). The S-module $\hat{S}$ has similar properties.

The conditions of the last two corollaries hold if the ring K is an equivalence situation, i.e., $I = R$ and $J = S$; such rings are considered in Sect. 3.8.

Below, we consider regular rings, as defined in Sect. 2.5.

Corollary 3.5.6 *Under the conditions of Corollary 3.5.4, the ring $Q(K)$ is regular if and only if $Q(R)$ and $Q(S)$ are regular rings.*

Proof The necessity of the conditions follows from Theorem 2.5.6.

Now we assume that $Q(R)$, $Q(S)$ are regular rings. We have the ring $T = \text{End}_K E$ represented in the form $(*)$ or $(**)$. It is sufficient to verify that this ring is semiprimitive; see Sect. 4.5 of the book [80]. This follows from Theorem 2.4.1 and the property that the modules $\hat{M}$ and $\hat{N}$ are reflexive with respect to $\hat{R}$ and $\hat{S}$, respectively. □

Closing this section, we briefly consider another extreme case, in which $(L(R, M),$ $L(N, S))$ is an essential submodule of $_K K$, where $K = \begin{pmatrix} R & M \\ N & S \end{pmatrix}$ is some formal matrix ring. In this case, the structure of the injective hull of the module $_K K$ is known; see Corollary 3.4.9. Then we can calculate the endomorphism ring of this hull and the maximal ring of fractions $Q(K)$. In the beginning of this section, it was noted that the above submodule is essential provided the trace ideals I, J are nilpotent. This situation is considered in the paper [99].

3.6 Flat Modules

In this section, we describe flat modules over formal matrix rings with zero trace ideals. We use right modules and right-side analogues of earlier obtained assertions and earlier defined constructions. The trace ideals I and J of the ring K often appear; see Sect. 2.1. We also note that isomorphisms, considered in the text, are canonical; this means that they act by a certain rule, which is easily specified.

In the study of flat modules, any information about tensor products is useful. We can calculate tensor products of K-modules with the use of tensor products of R-modules and tensor products of S-modules. Let $U = (C, D)$ be a right K-module and $V = (A, B)$ a left K-module.

Proposition 3.6.1 *There is an isomorphism of Abelian groups*

$$U \otimes_K V \cong (C \otimes_R A \oplus D \otimes_S B) / H,$$

where the subgroup H is generated by all elements of the form

$$c \otimes mb - cm \otimes b, \quad d \otimes na - dn \otimes a, \quad \text{where}$$
$$c \in C, \quad d \in D, \quad a \in A, \quad b \in B, \quad m \in M, \quad n \in N.$$

Proof The group $C \otimes_R A \oplus D \otimes_S B$ is isomorphic to the group $U \otimes_{R \times S} V$ under the correspondence of generator elements $c \otimes a + d \otimes b \to (c, d) \otimes (a, b)$. Set $G_1 = U \otimes_{R \times S} V$ and $G_2 = U \otimes_K V$. We define the tensor product as a factor group of a free group. Let F be the free Abelian group with basis consisting of all expressions

$$((c, d), (a, b)), \quad c \in C, \quad d \in D, \quad a \in A, \quad b \in B.$$

Then $G_1 = F/H_1$ and $G_2 = F/H_2$, where H_1, H_2 are the subgroups generated by elements of the known form. We point out the difference between these subgroups. The generator system of the group H_1 contains all elements of the form

$$((c, d), (ra, sb)) - ((cr, ds), (a, b)), \qquad r \in R, \quad s \in S;$$

the generator system of the group H_2 contains all elements of the form

$$((c, d), k(a, b)) - ((c, d)k, (a, b)), \qquad k = \begin{pmatrix} r & m \\ n & s \end{pmatrix} \in K,$$

and all remaining generators of these two groups coincide with each other. Therefore, $H_1 \subseteq H_2$. We have the relations

$$G_2 = F/H_2 \cong (F/H_1)/(H_2/H_1) = G_1/H, \qquad \text{where} \quad H = H_2/H_1.$$

The factor group H_2/H_1 is generated by the images of all generator elements of the group H_2, i.e., elements of the form

$$((c, d) \otimes (mb, na)) - ((dn, cm) \otimes (a, b)).$$

Using the above isomorphism, we obtain that $G_2 \cong (C \otimes_R A \oplus D \otimes_S B)/H$, where the (non-renamed) subgroup H is generated by all elements of the form $cm \otimes b + d \otimes na - dn \otimes a - cm \otimes b$; therefore, H is generated by all the above-specified elements. $\qquad\qquad\qquad\square$

Remarks Here are some general remarks and notations related to K-modules (A, B). We denote by L the ideal $\begin{pmatrix} I & IM \\ JN & J \end{pmatrix}$ of the ring K and set $\overline{K} = K/L$. We can identify the factor ring $\overline{K}$ with the matrix ring $\begin{pmatrix} R/I & M/IM \\ N/JN & S/J \end{pmatrix}$. We denote the last ring by $\begin{pmatrix} \overline{R} & \overline{M} \\ \overline{N} & \overline{S} \end{pmatrix}$. Since $\overline{MN} = 0 = \overline{NM}$, we have that $\overline{K}$ is a ring with zero trace ideals. The module (A, B) has submodules (IA, JB) and (MB, NA), and $(IA, JB) \subseteq (MB, NA)$. As was noted above, we can identify the factor modules $(A, B)/(IA, JB)$ and $(A, B)/(MB, NA)$ with the modules $(A/IA, B/JB)$ and $(A/MB, B/NA)$, respectively. Since $L(A, B) = (IA, JB)$, we have that $(A/IA, B/JB)$ and $(A/MB, B/NA)$ are $\overline{K}$-modules.

Now consider the ideal $L_1 = \begin{pmatrix} I & M \\ N & J \end{pmatrix}$ of the ring K. There exists an isomorphism

$$K/L_1 \cong R/I \times S/J = \overline{R} \times \overline{S}.$$

It follows from the relations $L_1(A, B) = (MB, NA)$ that $(A/MB, B/NA)$ is an $\overline{R} \times \overline{S}$-module.

Using the above notation, we formulate the following result.

Corollary 3.6.2 *Let (A, B) be a flat K-module.*

(1) $(A/IA, B/JB)$ *is a flat $\overline{K}$-module and $M/IM \otimes_S B/NA \cong MB/IA$, $N/JN \otimes_R A/MB \cong NA/JB$.*

(2) *A/MB is a flat $\overline{R}$-module and B/NA is a flat $\overline{S}$-module.*

(3) *If $I = 0$, then $M \otimes_S B/NA \cong MB$ and A/MB is a flat R-module, and for $N = 0$, we have $M \otimes_S B \cong MB$ and B is a flat S-module.*

Proof (1). We have the relation $(A/IA, B/JB) = (A, B)/L(A, B)$. It is known that $(A, B)/L(A, B)$ is a flat module. This follows, for example, from the Chase criterion; see [34, Proposition 11.33].

Set $\overline{A} = A/IA$ and $\overline{B} = B/JB$. Since $(\overline{A}, \overline{B})$ is a flat module, there exists an isomorphism

$$(0, \overline{M}) \otimes_{\overline{K}} (\overline{A}, \overline{B}) \cong (0, \overline{M})(\overline{A}, \overline{B}) = \overline{MB},$$

where $(0, \overline{M})$ is a right ideal of the ring $\overline{K}$. By Proposition 3.6.1, the tensor product from the left part is isomorphic to the factor group $(\overline{M} \otimes_{\overline{S}} \overline{B})/\overline{H}$ and the subgroup $\overline{H}$ is generated by the elements of the form $\overline{m} \otimes \overline{na}$ for all $\overline{m} \in \overline{M}$, $\overline{n} \in \overline{N}$, $\overline{a} \in \overline{A}$ (consider that $\overline{MN} = 0$). The group $\overline{H}$ is the image of the induced mapping $\overline{M} \otimes_{\overline{S}} \overline{NA} \to \overline{M} \otimes_{\overline{S}} \overline{B}$. Consequently, there is an isomorphism $(0, \overline{M}) \otimes_K (\overline{A}, \overline{B}) \cong \overline{M} \otimes_{\overline{S}} \overline{B/NA}$. Thus, we have the isomorphism

$$\overline{M} \otimes_{\overline{S}} \overline{B/NA} \cong \overline{MB}, \qquad \overline{m} \otimes (\overline{b} + \overline{NA}) \to \overline{mb}.$$

Under a more detailed representation, this isomorphism has the form $M/IM \otimes_S B/NA \cong MB/IA$. The second isomorphism is similarly proved.

(2). Similar to (1), it is proved that $(A/MB, B/NA)$ is a flat $\overline{R} \times \overline{S}$-module.

(3). The assertion directly follows from (1) and (2).   □

Remarks We apply to K-modules a standard procedure of transition from left modules to right modules. Let (X, Y) be a K-module and G an arbitrary Abelian group. The group of additive homomorphisms $\mathrm{Hom}((X, Y), G)$ is a right K-module with module multiplication defined by the relation

$$(\eta k)(x, y) = \eta(k(x, y)), \eta \in \mathrm{Hom}((X, Y), G), k \in K, x \in X, y \in Y.$$

Similarly, the group $\mathrm{Hom}(X, G)$ (resp., $\mathrm{Hom}(Y, G)$) is a right R-module (resp., a right S-module). We can consider the group of row vectors $(\mathrm{Hom}(X, G), \mathrm{Hom}(Y, G))$ as a right K-module. The module multiplication is defined by the relations

$$(\alpha m)y = \alpha(my), \qquad (\beta n)x = \beta(nx), \qquad \text{where}$$
$$\alpha \in \text{Hom}(X, G), \beta \in \text{Hom}(Y, G), m \in M, n \in N, x \in X, y \in Y.$$

There exists a canonical K-module isomorphism

$$\text{Hom}((X, Y), G) \to (\text{Hom}(X, G), \text{Hom}(Y, G)),$$
$$\text{Hom}((X, Y), G) \ni \eta \to (\eta|_X, \eta|_Y) \in (\text{Hom}(X, G), \text{Hom}(Y, G)).$$

With the use of this isomorphism, we identify the K-modules $\text{Hom}((X, Y), G)$ and $(\text{Hom}(X, G), \text{Hom}(Y, G))$.

If V is a module over some ring T, then the right T-module $\text{Hom}(V, \mathbb{Q}/\mathbb{Z})$ is called the *character module* of the module V; it is denoted by V^*. The character module of a K-module (X, Y) is (X^*, Y^*). It is well-known that a T-module V is flat if and only if the character module V^* is injective. Since we have a description of injective K-modules, we can conditionally assume that we have a description of flat K-modules. We note that this description is difficult to formulate in terms of the initial module (X, Y). However, it is sometimes possible. For example, it is easy to obtain the following result (cf. Proposition 3.4.1).

Proposition 3.6.3 *A K-module $(X, T(X))$ is flat if and only if X is a flat R-module. A similar assertion holds for an S-module Y and the K-module $(T(Y), Y)$. Thus, the functor T from Sect. 3.1 preserves flat modules.*

Proof We identify the character module of the module $(X, T(X))$ with the right K-module $(X^*, T(X)^*)$. There are natural isomorphisms of right S-modules

$$T(X)^* \cong \text{Hom}_{\mathbb{Z}}(N \otimes_R X, \mathbb{Q}/\mathbb{Z}) \cong \text{Hom}_R(N, \text{Hom}(X, \mathbb{Q}/\mathbb{Z})) = \text{Hom}_R(N, X^*).$$

Thus, the character module of the K-module $(X, T(X))$ coincides with $(X^*, \text{Hom}_R (N, X^*))$. By the right-side analogue of Proposition 3.4.1, the module $(X^*, \text{Hom}_R (N, X^*))$ is injective if and only if the module X^* is injective. The last condition is equivalent to the property that the module X is flat.

The case of the module Y is similar. $\square$

Remark It directly follows from Proposition 3.6.3 that the module $(A, 0)$ (resp., $(0, B)$) is flat if and only if A is a flat module and $N \otimes_R A = 0$ (resp., B is a flat module and $M \otimes_S B = 0$). In addition, Theorem 3.4.2 is equivalent to the following result.

Corollary 3.6.4 *Let (A, B) be a K-module, $N A = B$ and $M B = A$. Then the module (A, B) is flat if and only if A, B are flat modules.*

Proof The character module of the module (A, B) is equal to (A^*, B^*). We have $(\eta m)b = \eta(mb)$ for all $\eta \in A^*$, $m \in M$, and $b \in B$. Therefore,

$$L(A^*) = \text{Hom}(A/MB, \mathbb{Q}/\mathbb{Z}) = (A/MB)^*,$$

where the module $\text{Hom}(A/MB, \mathbb{Q}/\mathbb{Z})$ is identified with the set of all $\eta: A \to \mathbb{Q}/\mathbb{Z}$ with $\eta(MB) = 0$. In addition, $L(B^*) = (B/NB)^*$. Therefore, it follows from the assumption that $L(A^*) = 0 = L(B^*)$. The module (A^*, B^*) satisfies the conditions of the right-side variant of Theorem 3.4.2. Consequently, the module (A^*, B^*) is injective if and only if A^*, B^* are injective modules. The last property is equivalent to the property that A, B are flat modules. $\qquad\square$

Under an additional condition that the trace ideals of the ring K are equal to the zero ideal, we obtain the following result.

Theorem 3.6.5 *If K is a ring with zero trace ideals, then the K-module (A, B) is flat if and only if A/MB is a flat R-module, B/NA is a flat S-module and there are isomorphisms*

$$M \otimes_S B/NA \cong MB, \qquad N \otimes_R A/MB \cong NA.$$

Proof If (A, B) is a flat K-module, then it follows from Corollary 3.6.2(3) that A/MB is a flat R-module, B/NA is a flat S-module, $M \otimes_S B/NA \cong MB$ and $N \otimes_R A/MB \cong NA$.

We prove that the conditions are sufficient. We have the K-modules $(MB, 0)$ and (A, NA). We consider the factor module $(A, NA)/(MB, 0)$, which coincides with the module $(A/MB, NA)$. Then $M(NA) = 0$ and $n(a + MB) = na$ for all $n \in N$ and $a \in A$. By assumption, the S-modules $N \otimes_R A/MB$ and NA are isomorphic. This isomorphism is induced by the correspondence $n \otimes (a + MB) \to na$ of generator elements. By Lemma 3.1.2, the identity mapping of the module A/MB induces the K-module homomorphism $(A/MB, T(A/MB)) \to (A/MB, NA)$, which coincides with the above isomorphism in the second positions. Now it follows from Proposition 3.6.3 and the conditions of the theorem that $(A/MB, NA)$ is a flat module. Similarly, it is proved that $(MB, B/NA)$ is a flat module. Therefore, the character modules $(A/MB, NA)^*$ and $(MB, B/NA)^*$ are injective.

Now we prove that the direct sum of these two modules is isomorphic to the module $(A, B)^*$. Thus, we have to verify that there exists an isomorphism of right K-modules

$$\left((A/MB)^* \oplus (MB)^*, (NA)^* \oplus (B/NA)^*\right) \cong \left(A^*, B^*\right). \tag{3.6.1}$$

We use the following relations from the proof of Corollary 3.6.4:

$$(A/MB)^*M = 0 = (B/NA)^*N. \tag{3.6.2}$$

Since the $\mathbb{Z}$-module $\mathbb{Q}/\mathbb{Z}$ is injective, there exists an exact sequence of right R-modules

$$0 \to (A/MB)^* \to A^* \xrightarrow{\pi} (MB)^*. \tag{3.6.3}$$

It has already been mentioned that we identify the module $(A/MB)^*$ with its image in A^*, which consists of all homomorphisms $A \to \mathbb{Q}/\mathbb{Z}$ annihilating MB. In addition, we note that the mapping π maps an arbitrary homomorphism $A \to \mathbb{Q}/\mathbb{Z}$ onto its restriction to MB. The R-module $(A/MB)^*$ is injective, since A/MB is a flat module. Consequently, there exists a direct decomposition $A^* = (A/MB)^* \oplus V$ with summand V which is isomorphic to the module MB^*. To obtain the required isomorphism (3.6.1), it is necessary to choose in some way a module V and an isomorphism between $(MB)^*$ and V. Since the sequence (3.6.3) splits, there exists a monomorphism $\varepsilon \colon (MB)^* \to A^*$, for which $\varepsilon\pi$ is the identity mapping. We take Im ε as V, and we take ε as the isomorphism between $(MB)^*$ and V. As a result, we obtain an isomorphism of right K-modules

$$\Phi \colon (A/MB)^* \oplus (MB)^* \to A^*,$$

which acts identically on $(A/MB)^*$, and $\Phi(\alpha)|_{MB} = \alpha$ for all $\alpha \in (MB)^*$. We can also find an isomorphism $\Psi \colon (NA)^* \oplus (B/NA)^* \to B^*$ with similar properties.

We prove that the pair (Φ, Ψ) defines the isomorphism (3.6.1). It is sufficient to verify the validity of the right-side analogues of the two relations mentioned in Sect. 3.1. Namely, we verify that

$$\Phi(\beta n) = \Psi(\beta)n, \quad \Psi(\alpha m) = \Phi(\alpha)m \qquad \text{for all}$$
$$\beta \in (NA)^* \oplus (B/NA)^*, \quad \alpha \in (A/MB)^* \oplus (MB)^*, \quad n \in N, \quad m \in M.$$

To do this, we consider the relations (3.6.2) and the choice of isomorphisms Φ and Ψ. We take concrete β, n and write $\beta = \gamma + \delta$, where $\gamma \in (NA)^*$, $\delta \in (B/NA)^*$. Since $\delta n = 0$, we have $\beta n = \gamma n \in (A/MB)^*$. For any element $a \in A$ from the definition of the mapping Φ, we have $\Phi(\beta n) = \Phi(\gamma n)a$. On the other hand, it follows from the definition of the mapping Ψ that

$$(\Psi(\delta)n)a = (\Psi(\delta))(na) = 0, \qquad (\Psi(\beta)n)a = (\Psi(\gamma)n)a = (\Psi(\gamma))(na).$$

Since $\gamma(na) = (\gamma n)a$, we have $\Phi(\beta n) = \Psi(\beta)n$. The second relation is similarly verified. Thus, the module $(A, B)^*$ is injective. Therefore, (A, B) is a flat module. $\qquad\square$

Corollary 3.6.6 ([35]) *A module (A, B) over the ring $\begin{pmatrix} R & M \\ 0 & S \end{pmatrix}$ is flat if and only if A/MB, B are flat modules and $M \otimes_S B \cong MB$.*

3.7 Projective and Hereditary Modules and Rings

We describe projective modules and hereditary modules over a formal matrix ring K with zero trace ideals. Then we apply this description to look for conditions under which the ring K is hereditary.

As above, I and J denote the trace ideals of the ring K. In this section, we usually denote K-modules by (P, Q).

Our first result can be proved using an argument similar to that of the proof of Proposition 3.4.1. We only have to cite Lemma 3.1.2 instead of Lemma 3.1.3.

Proposition 3.7.1 *If X is a projective R-module and Y is a projective S-module, then $(X, T(X))$ and $(T(Y), Y)$ are projective K-modules. The converse also holds.*

It follows from Proposition 3.7.1 that the K-module $(X, 0)$ is projective if and only if X is a projective R-module and $N \otimes_R X = 0$. Similarly, we obtain that the assertion holds for the K-module $(0, Y)$.

In Sect. 3.6, we defined the ideals L and L_1 of the ring K.

Corollary 3.7.2 *Let (P, Q) be a projective K-module.*

(1) $(P/IP, Q/JQ)$ *is a projective K/L-module.*
(2) P/MQ *is a projective R/I-module and Q/NP is a projective S/J-module.*

Proof If V is a projective module over some ring T and A is an ideal in T, then V/AV is a projective T/A-module. Using this property for $T = K$, $A = L$ and $T = K$, $A = L_1$, it is easy to prove assertions (1) and (2). $\square$

Theorem 3.7.3 *Let K be a ring with zero trace ideals and (P, Q) a K-module. Then the following conditions are equivalent.*

(1) (P, Q) *is a projective module.*
(2) P/MQ *is a projective R-module, Q/NP is a projective S-module, $M \otimes_S Q/NP \cong MQ$ and $N \otimes_R P/MQ \cong NP$.*
(3) *There exist a projective R-module X and a projective S-module Y such that $(P, Q) = (X, NP) \oplus (MQ, Y)$, $M \otimes_S Y \cong MQ$ and $N \otimes_R X \cong NP$.*
(4) *There exist a projective R-module X and a projective S-module Y such that $(P, Q) \cong (X, T(X)) \oplus (T(Y), Y)$.*

Proof (1) $\Rightarrow$ (2). By Corollary 3.7.2, P/MQ is a projective R-module and Q/NP is a projective S-module. By Corollary 3.6.2, $M \otimes_S Q/NP \cong MQ$ and $N \otimes_R P/MQ \cong NP$.

(2) $\Rightarrow$ (3). First, we make one remark. Since $Y \subseteq Q$, there exists an induced homomorphism $M \otimes_S Y \to M \otimes_S Q$. The composition of this homomorphism and the homomorphism of module multiplication $M \otimes_S Q \to MQ$ provides the homomorphism $M \otimes_S Y \to MQ$. In condition (3), it is assumed that this homomorphism is an isomorphism. We have $P = X \oplus MQ$ and $Q = NP \oplus Y$, where $X \cong P/MQ$ and $Y \cong Q/NP$. Since (X, NP) and (MQ, Y) are K-modules, we obtain the relations

$$(P, Q) = (X \oplus MQ, NP \oplus Y) = (X, NP) \oplus (MQ, Y).$$

$(3) \Rightarrow (4)$. The assertion follows from Lemma 3.1.2 and the property that

$$(X, T(X)) \cong (X, NP), \qquad (T(Y), Y) \cong (MQ, Y).$$

The implication $(4) \Rightarrow (1)$ follows from Proposition 3.7.1.   □

For the ring of triangular matrices, we obtain the following result; the equivalence $(1) \Leftrightarrow (2)$ of this result is proved in [52].

Corollary 3.7.4 *Let* $K = \begin{pmatrix} R & M \\ 0 & S \end{pmatrix}$ *and* (P, Q) *a* K*-module. Then the following conditions are equivalent.*

(1) (P, Q) *is a projective module.*
(2) P/MQ *is a projective* R*-module,* Q *is a projective* S*-module and* $M \otimes_S Q \cong MQ$.
(3) Q *is a projective* S*-module,* $M \otimes_S Q \cong MQ$ *and there exists a projective* R*-module* X *with* $(P, Q) = (X, 0) \oplus (MQ, Q)$.
(4) Q *is a projective* S*-module and there exists a projective* R*-module* X *with* $(P, Q) \cong (X, 0) \oplus (T(Q), Q)$.

A module is said to be *hereditary* if all its submodules are projective. We begin the study of hereditary modules with the following results.

Proposition 3.7.5 *If* (P, Q) *is a hereditary module, then* P *and* Q *are hereditary modules.*

Proof As always, we prove the assertion for one of the modules P or Q, since for the second module, the assertion is similarly proved. We take an arbitrary submodule A in P. By assumption, (A, NP) is a projective submodule of (P, Q). Let $\{a_t \mid t \in T\}$ be some generator system of the R-module A. Then $\{(a_t, 0) \mid t \in T\}$ is a generator system of the K-module (A, NA). By the Dual Basis Lemma [34, Lemma 3.23], there exist K-module homomorphisms $F_t : (A, NA) \to K$, $t \in T$ such that every element $v \in (A, NA)$ is equal to $\sum_{t \in T} F_t(v)(a_t, 0)$, where almost all elements $F_t(v)$ are equal to the zero. We denote by h the additive homomorphism

$$K \to R, \qquad \begin{pmatrix} r & * \\ * & * \end{pmatrix} \to r;$$

$*$ denotes elements which are not important for us. For every $t \in T$, there exists an additive homomorphism $f_t : A \to R$ such that $f_t(a) = (F_t h)(a, 0)$, $a \in A$. We verify that $f_t(ra) = rf_t(a)$ for all $r \in R$ and $a \in A$. We have

$$f_t(ra) = h(F_t(ra, 0)) = h(r F_t(a, 0)) = h\left(r \begin{pmatrix} c & * \\ * & * \end{pmatrix}\right) = h\begin{pmatrix} rc & * \\ * & * \end{pmatrix} = rc,$$

$$rf_t(a) = r(h(F_t(a, 0))) = rh\begin{pmatrix} c & * \\ * & * \end{pmatrix} = rc.$$

Consequently, all f_t are R-module homomorphisms. For every element $a \in A$, we have

$$(a, 0) = \sum F_t(a, 0)(a_t, 0) = \sum \begin{pmatrix} r_t & * \\ * & * \end{pmatrix} (a_t, 0) = \sum (r_t a_t, *), \quad a = \sum r_t a_t,$$

where the subscript $t \in T$ is omitted everywhere, $r_t \in R$ and almost all r_t are equal to zero. Thus, $a = \sum f_t(a)a_t$. By the Dual Basis Lemma, the R-module A is projective. Consequently, P is a hereditary module. □

By applying Theorem 3.7.3, we obtain the description of hereditary K-modules.

Theorem 3.7.6 *Let K be a ring with zero trace ideals and (P, Q) a K-module. The module (P, Q) is hereditary if and only if the following conditions hold.*

(1) *P and Q are hereditary modules.*
(2) *For any submodule B in Q, the module P/MB is projective and $M \otimes_S B \cong MB$.*
(3) *For any submodule A in P, the module Q/NA is projective and $N \otimes_R A \cong NA$.*

Proof We assume that (P, Q) is a hereditary module. By Proposition 3.7.5, P and Q are hereditary modules. We take some submodule B of Q and the K-module (MB, B). The module (MB, B) is projective, since it is a submodule of the hereditary module (P, Q). It follows from Theorem 3.7.3 that $M \otimes_S B \cong MB$. Now we take the submodule $(P, B + NP)$ of the module (P, Q). By Theorem 3.7.3, the module P/MB is projective. Similar arguments hold for any submodule A of P.

Now we assume that the conditions of the theorem hold. Let (A, B) be some submodule of (P, Q). We have the relations

$$P = X \oplus MB, \quad M \otimes_S B \cong MB, \quad Q = Y \oplus NA, \quad N \otimes_R A \cong NA,$$

where X and Y are some projective modules. Since $MB \subseteq A$ and $NA \subseteq B$, there exist decompositions

$$A = (A \cap X) \oplus MB, \qquad B = (B \cap Y) \oplus NA,$$
$$(A, B) = (A \cap X, NA) \oplus (MB, B \cap Y).$$

We verify that the last decomposition satisfies the conditions of Item (3) of Theorem 3.7.3. Indeed, $A \cap X$ and $B \cap Y$ are projective modules. Then we have

$$MB \cong M \otimes_S B \cong M \otimes_S (B \cap Y) \oplus M \otimes_S NA.$$

However, $M \otimes_S NA \cong MNA = 0$. Therefore, $M \otimes_S (B \cap Y) \cong MB$. Similarly, $N \otimes_S (A \cap X) \cong NA$. By Theorem 3.7.3, the module (A, B) is projective. Therefore, (P, Q) is a hereditary module. □

Corollary 3.7.7 *A module* (P, Q) *over the ring of triangular matrices* $K = \begin{pmatrix} R & M \\ 0 & S \end{pmatrix}$ *is hereditary if and only if* P *and* Q *are hereditary modules,* P/MB *is a projective module for any submodule* B *in* Q, *and* $M \otimes_S B \cong MB$.

Remark Assertions similar to the results of Sects. 3.2–3.7 hold for right K-modules (for example, analogues of Theorems 3.7.3 and 3.7.6 hold). See the beginning of Sect. 3.1 for details.

A ring T is said to be *left hereditary* (resp., *right hereditary*) if T is a left (resp., right) hereditary T-module, i.e., every left (resp., right) ideal of the ring T is a projective left (resp., right) T-module.

We apply Theorem 3.7.6 to the ring K.

Corollary 3.7.8 *A formal matrix ring* K *with zero trace ideals is left hereditary if and only if the following conditions hold.*

(1) *The rings* R *and* S *are left hereditary.*
(2) M *is a flat* S-module, N *is a flat* R-module, *and* $M \otimes_S N = 0 = N \otimes_R M$.
(3) M/ML *is a projective* R-module for any left ideal L of the ring S.
(4) N/NL *is a projective* S-module for any left ideal L of the ring R.

Proof We remark that the ring K is left hereditary if and only if (R, N) and (M, S) are hereditary left K-modules.

Let (R, N) be a hereditary K-module. By Theorem 3.7.6, the ring R is left hereditary. In addition, for any left ideal L of the ring R, the S-module N/NL is projective and the canonical homomorphism $N \otimes_R L \to NL$ is an isomorphism. The last property is equivalent to the property that N is a flat R-module. Thus, $M \otimes_S N = 0 = N \otimes_R M$. With the use of the hereditary K-module (M, S), the remaining conditions are similarly verified.

We assume that conditions (1)–(4) hold. It follows from Theorem 3.7.6 that the left K-modules (R, N) and (M, S) are hereditary. We just make a few clarifications. Since N is a flat S-module, $M \otimes_S B \subseteq M \otimes_S N = 0$ for any S-submodule B in N. Since $MB = 0$, we have $M \otimes_S B = MB$. In addition, $N \otimes_R L \cong NL$ for any left ideal L of the ring R, since N is a flat R-module. Similarly, it is proved that (M, S) is a hereditary module. $\qquad\square$

We have three corollaries for the ring of triangular matrices. The first corollary is used in Sect. 3.9 in the study of Abelian groups with hereditary endomorphism rings.

Corollary 3.7.9 ([40]) *A ring* $\begin{pmatrix} R & M \\ 0 & S \end{pmatrix}$ *is left hereditary if and only if* R *and* S *are left hereditary rings,* M *is a flat* S-module and M/ML *is a projective* R-module for any left ideal L of the ring S.

Corollary 3.7.10 *If* R, S *are Artinian semiprimitive rings, then* $\begin{pmatrix} R & M \\ 0 & S \end{pmatrix}$ *is a left and right hereditary ring for every* R-S-bimodule M.

Corollary 3.7.11 *A ring* $\begin{pmatrix} R & R \\ 0 & R \end{pmatrix}$ *is left (or right) hereditary if and only if R is an Artinian semiprimitive ring.*

Proof First, we consider the case of left hereditary rings. If R is an Artinian semiprimitive ring, then the ring $\begin{pmatrix} R & R \\ 0 & R \end{pmatrix}$ is left hereditary by Corollary 3.7.10.

We assume that the ring $\begin{pmatrix} R & R \\ 0 & R \end{pmatrix}$ is left hereditary. By Corollary 3.7.9, the module $R/RL = R/L$ is projective for any left ideal L of the ring R. Therefore, L is a direct summand of the module $_R R$. Therefore, R is an Artinian semiprimitive ring.

The case of right hereditary rings can be proved via a transition to the opposite ring $\begin{pmatrix} R^\circ & 0 \\ R^\circ & R^\circ \end{pmatrix}$ (such rings are mentioned in Sect. 2.1). □

The following well-known fact follows from Corollary 3.7.11.

For any division ring D, the ring $\begin{pmatrix} D & D \\ 0 & D \end{pmatrix}$ is hereditary. The following result can be considered as a generalization of this fact.

Corollary 3.7.12 *Let D, F be two division rings and* $K = \begin{pmatrix} D & V \\ W & F \end{pmatrix}$ *a formal matrix ring. The ring K is left (or right) hereditary if and only if* $D \cong F$, *V and W are one-dimensional D-spaces and F-spaces, and either K is not a ring with zero trace ideals, or K is a ring of (upper or lower) triangular matrices.*

Proof Let K be a left or right hereditary ring. Since D and F are division rings, only the following three possibilities are possible for the trace ideals I and J of the ring K:

(1) $I = 0$; (2) $I = D$ and $J = F$; (3) $J = 0$.
By considering the products VWV and WVW, it is easy to verify that either $I = D$ and $J = F$, or $I = 0 = J$. In the first case, we have that V and W are one-dimensional D-spaces and F-spaces and $D \cong \mathrm{End}_F(V)$ by Lemma 3.8.3. For $I = 0 = J$, it follows from Corollary 3.7.8 that $V \otimes_F W = 0$; therefore, $V = 0$ or $W = 0$.

Conversely, if the formal matrix ring $\begin{pmatrix} D & D \\ D & D \end{pmatrix}$ is not a ring with zero trace ideals, then it is isomorphic to the "ordinary" ring of 2×2 matrices over D. The case of the ring of triangular matrices is contained in Corollaries 3.7.10 and 3.7.11. □

A ring T is said to be *left perfect* if each left T-module has a projective hull. A ring T is left perfect if and only if every flat left T-module is projective.

Corollary 3.7.13 *If a ring* $K = \begin{pmatrix} R & M \\ N & S \end{pmatrix}$ *is left perfect, then the rings R and S are left perfect. For a ring with zero trace ideals the converse holds.*

Proof Let the ring K be left perfect. We take an arbitrary flat R-module X. Then $(X, T(X))$ is a flat module; see Proposition 3.6.3. Since K is left perfect, the module $(X, T(X))$ is projective. By Proposition 3.7.1, the module X is projective. Therefore, the ring R is left perfect. It is similarly proved that the ring S is left perfect.

Conversely, we assume that the rings R and S are left perfect. By Theorems 3.6.5 and 3.7.3, each flat left K-module is projective. Therefore, K is left perfect. □

Remark In general, it is not known whether the converse to Corollary 3.7.13 holds.

Now we can offer a satisfactory description of flat modules from Theorem 3.6.5.

Theorem 3.7.14 *Let K be a ring with zero trace ideals. Then for any flat K-module (P, Q) there exist a flat R-module X and a flat S-module Y such that $(P, Q) \cong (X, T(X)) \oplus (T(Y), Y)$.*

Proof The module (P, Q) is a direct limit of finitely generated projective K-modules $\{(A_i, B_i), i \in I; (e_i^j, f_i^j)\}$, where $(e_i^j, f_i^j): (A_i, B_i) \longrightarrow (A_j, B_j), i \leqslant j$ [34, Chap. 5, Ex. 55.9]. It follows from Theorem 3.7.3 that for every i, there are a projective R-module X_i and a projective S-module Y_i such that $(A_i, B_i) \cong (X_i, T(X_i)) \oplus (T(Y_i), Y_i), X_i = A_i/MB_i$ and $Y_i = B_i/NA_i$. It follows from Proposition 3.2.1 that X_i and Y_i are finitely generated modules. Since $e_i^j(MB_i) \subseteq MB_j$ and $f_i^j(NA_i) \subseteq NA_j$, the homomorphisms e_i^j and f_i^j induce the homomorphisms $X_i \longrightarrow X_j$ and $Y_i \longrightarrow Y_j$, respectively. We denote them by the same symbols e_i^j and f_i^j.

We can verify that there exist direct spectra $\{X_i, i \in I; e_i^j\}$, $\{Y_i, i \in I; f_i^j\}$, $\{(X_i, T(X_i)), i \in I; (e_i^j, 1 \otimes e_i^j)\}$ and $\{(T(Y_i), Y_i), i \in I; (1 \otimes f_i^j, f_i^j)\}$. Using the definition of a direct limit, we can directly verify that

$$\lim_{\to I} (X_i, T(X_i)) \cong (\lim_{\to I} X_i, T(\lim_{\to I} X_i));$$

and similar assertions hold for the other spectra. (Note that direct limits commute with tensor product). Set

$$X = \lim_{\to I} X_i, \quad Y = \lim_{\to I} Y_i.$$

Then X and Y are flat modules and we have isomorphisms

$$(P, Q) \cong \lim_{\to I} (A_i, B_i)$$

$$\cong \lim_{\to I} (X_i, T(X_i)) \oplus \lim_{\to I} (T(Y_i), Y_i) \cong (X, T(X)) \oplus (T(Y), Y). \quad \square$$

3.8 Equivalences Between the Categories R-Mod, S-Mod, and K-Mod

It is natural to specially consider formal matrix rings $K = \begin{pmatrix} R & M \\ N & S \end{pmatrix}$ such that $I = R$ and $J = S$. This case is opposite to the case $I = 0 = J$, where K is a ring with zero trace ideals. Such rings appear in the study of equivalences between categories of R-modules and S-modules. The main results in this field are known as Morita theorems. Sections 21 and 22 of the book [7] contain a detailed exposition of various questions related to equivalences of module categories. Here we describe the role of the ring K in the study of equivalences of categories. It becomes clear that the structure of modules over the ring K for $I = R$ and $J = S$ does not depend on the bimodules M and N; it is determined by the structure of the corresponding R-modules or S-modules.

Let $K = \begin{pmatrix} R & M \\ N & S \end{pmatrix}$ be a formal matrix ring, $\varphi: M \otimes_S N \to R$ and $\psi: N \otimes_R M \to S$ the bimodule homomorphisms defined in Sect. 2.1. The images I and J of these homomorphisms are called the *trace ideals* of the ring K.

We recall that a given K-module (A, B) is associated with two pairs of the homomorphisms of module multiplication

$$f: M \otimes_S B \to A, \qquad g: N \otimes_R A \to B,$$
$$f': B \to \mathrm{Hom}_R(M, A), \qquad g': A \to \mathrm{Hom}_S(N, B);$$

see the beginning of Sect. 3.1.

Lemma 3.8.1 *If $I = R$ and $J = S$, then f, g, f', g' are isomorphisms.*

Proof We can represent

$$1 = m_1 n_1 + \ldots + m_k n_k, \qquad \text{where} \quad m_i \in M, \quad n_i \in N, \quad i = 1, \ldots, k.$$

Since $A = IA \subseteq MB$, we have that f is surjective. We assume that $f(x_1 \otimes b_1 + \ldots + x_\ell \otimes b_\ell) = 0$ for some $x_j \in M$, $b_j \in B$, $j = 1, \ldots, \ell$. We have

$$\sum_j x_j \otimes b_j = \sum_{i,j} (m_i n_i)(x_j \otimes b_j) = \sum_{i,j} m_i (n_i x_j) \otimes b_j$$
$$= \sum_{i,j} m_i \otimes n_i (x_j b_j) = \sum_i (m_i \otimes n_i) \cdot \sum_j x_j b_j = \sum_i (m_i \otimes n_i) \cdot 0 = 0.$$

Consequently, f is an isomorphism.

Similarly, it follows from the relation $J = S$ that

$$1 = n_1 m_1 + \ldots + n_k m_k, \qquad \text{where} \quad n_i \in N, \quad m_i \in M, \quad i = 1, \ldots, k;$$

without risk of confusion, we use the same symbols n_i, m_i. Similarly, we obtain that g is an isomorphism.

If $f'(b) = 0$, then $b = \sum_i (n_i m_i)b = \sum_i n_i (m_i b) = 0$, since $m_i b = 0$. Let α be an arbitrary homomorphism $M \to A$. For any $m \in M$, we have

$$\alpha(m) = \alpha\left(m \sum_i n_i m_i \right) = \alpha\left(\sum_i (mn_i)m_i \right)$$

$$= \sum_i (mn_i)\alpha(m_i) = m\left(\sum_i n_i \alpha(m_i) \right).$$

Therefore, $\alpha = f'\left(\sum_i n_i \alpha(m_i) \right)$. Thus, f' is an isomorphism. Similarly, we obtain that g' is an isomorphism. $\qquad\square$

Corollary 3.8.2 *Let $I = R$, $J = S$ and (A, B) a K-module.*

(1) *$MB = A$, $NA = B$, and we have canonical K-module isomorphisms $(A, T(A)) \cong (A, B)$, $(T(B), B) \cong (A, B)$, $(A, B) \cong (A, H(A))$, $(A, B) \cong (H(B), B)$.*
(2) *There are canonical ring isomorphisms*
   $\mathrm{End}_R A \cong \mathrm{End}_K(A, B) \cong \mathrm{End}_S B$.

Proof In (1), the required isomorphisms are $(1, g)$, $(f, 1)$, $(1, f')$, $(g', 1)$, respectively; see the remarks after Corollary 3.1.4. The assertion (2) follows from Corollary 3.1.4. $\qquad\square$

We recall several notions from module theory. Let C be an R-S-bimodule. For every element $r \in R$, the mapping $\alpha_r : c \to rc$, $c \in C$, is an S-homomorphism; it is called the *homothety* of the R-module C with coefficient r. The ring homomorphism $R \to \mathrm{End}_S C$, $r \to \alpha_r$, is called the *homothety mapping*. There is another homothety mapping, namely, $S \to \mathrm{End}_R C$, $s \to \beta_s$, where $\beta_s(c) = cs$, $s \in S$, $c \in C$. For example, let C be an R-module and $S = \mathrm{End}_R C$. Then C is an R-S-bimodule. Consequently, there exists a homothety mapping $R \to \mathrm{End}_S C$. Here $\mathrm{End}_S C$ is the biendomorphism ring of the R-module C.

In what follows, G^n denotes the direct sum of n isomorphic copies of the module G.

A module G over a ring T is called a *generator* or a *generator module* if one of the following equivalent conditions holds.

(1) The sum of images of each of the homomorphisms from G into R coincides with R.
(2) For any T-module X, the sum of images of each of the homomorphisms from G into X coincides with X.
(3) For all T-module homomorphisms $\alpha : G \to X$ and $\beta, \gamma : X \to Y$ such that $\alpha\beta = \alpha\gamma$, we have the relation $\beta = \gamma$.
(4) There exist a positive integer n and a T-module H such that $G^n \cong T \oplus H$.

A finitely generated projective generator is also called a *progenerator* or a *pro-generator module*.

We apply Lemma 3.8.1 to the ring K. First, we remark that each of the K-modules (R, N) and (M, S) provides four homomorphisms of module multiplication. In addition to the familiar homomorphisms, $\varphi \colon M \otimes_S N \to R$ and $\psi \colon N \otimes_R M \to S$, there are homomorphisms $\varphi' \colon N \to \operatorname{Hom}_R(M, R)$ and $\psi' \colon M \to \operatorname{Hom}_S(N, S)$, canonical isomorphisms $N \otimes_R R \to N$, $M \otimes_S S \to M$, and two homothety mappings $R \to \operatorname{End}_S N$, $S \to \operatorname{End}_R M$. We find that $\varphi'(n)(m) = mn$, $n \in N$, $m \in M$; the homomorphism ψ' acts similarly. Among the eight homomorphisms of module multiplication for the right K-modules (R, M) and (N, S), there are homomorphisms $N \to \operatorname{Hom}_S(M, S)$, $M \to \operatorname{Hom}_R(N, R)$, and two homothety mappings $R \to \operatorname{End}_S M$, $S \to \operatorname{End}_R N$.

Lemma 3.8.3 *Let K be a formal matrix ring such that $I = R$ and $J = S$.*

(1) *All the above-mentioned 16 bimodule homomorphisms are isomorphisms.*
(2) *Each of the modules $_R M$, M_S, $_S N$, and N_R is a progenerator.*

Proof (1). The assertion is a special case of Lemma 3.8.1.
(2). As in Lemma 3.8.1, let

$$1 = m_1 n_1 + \ldots + m_k n_k, \quad \text{where} \quad m_i \in M, \quad n_i \in N, \quad i = 1, \ldots, k.$$

Set $\alpha_i = \varphi'(n_i)$, $i = 1, \ldots, k$, and consider the homomorphism $\gamma = \alpha_1 + \ldots + \alpha_k \colon M^k \to R$. Since $\gamma(m_1 + \ldots + m_k) = m_1 n_1 + \ldots + m_k n_k = 1$, we have that γ is an epimorphism onto the projective module R. Therefore, γ splits and $M^k \cong R \oplus X$ for some module X. Thus, M is a generator. We can repeat this argument for remaining three modules. In particular, there exists an isomorphism $M^k \cong S \oplus Y$ for some right S-module Y. From isomorphisms of left R-modules

$$R^k \cong \operatorname{Hom}_S(M, M)^k \cong \operatorname{Hom}_S(M^k, M)$$
$$\cong \operatorname{Hom}_S(S \oplus Y, M) \cong \operatorname{Hom}_S(S, M) \oplus \operatorname{Hom}_S(Y, M) \cong M \oplus X,$$

it follows that M is a finitely generated projective R-module. We can repeat this argument for the remaining modules. $\qquad\square$

If there exists a formal matrix ring $\begin{pmatrix} R & M \\ N & S \end{pmatrix}$, then we can consider a so-called a *pre-equivalence situation* or a *Morita context* $(R, S, M, N, \varphi, \psi)$, where R and S are rings, $_R M_S$ and $_S N_R$ are bimodules, $\varphi \colon M \otimes_S N \to R$ and $\psi \colon N \otimes_R M \to S$ are bimodule homomorphisms, and the associativity relations

$$(mn)m' = m(nm'), \quad (nm)n' = n(mn'), \quad \forall m, m' \in M, \ n, n' \in N$$

hold. There exists an obvious one-to-one correspondence between formal matrix rings and pre-equivalence situations. Therefore, it is convenient to call the ring

$\begin{pmatrix} R & M \\ N & S \end{pmatrix}$ a *pre-equivalence situation* or a *Morita context*. If φ and ψ are isomorphisms, then $(R, S, M, N, \varphi, \psi)$ or the ring $\begin{pmatrix} R & M \\ N & S \end{pmatrix}$ is called an *equivalence situation*.

Remark We verify that it is possible to obtain a "standard" pre-equivalence situation if we start with an arbitrary module (for convenience, we actually repeat Example 1 from Sect. 2.2). Let M be a module over some ring R. We denote by S the endomorphism ring of the R-module M. Then M is an R-S-bimodule. Then we set $M^* = \operatorname{Hom}_R(M, R)$. The group M^* is an S-R-bimodule, where

$$(s\alpha)m = \alpha(s(m)), \quad (\alpha r)m = \alpha(mr), \quad \alpha \in M^*, \ s \in S, \ r \in R, \ m \in M.$$

There exist an R-R-bimodule homomorphism $\varphi: M \otimes_S M^* \to R$ and an S-S-bimodule homomorphism $\psi: M^* \otimes_R M \to S$ defined by the relations

$$\varphi\left(\sum m_i \otimes \alpha_i\right) = \sum \alpha_i(m_i), \quad \left(\psi\left(\sum \alpha_i \otimes m_i\right)\right)(m) = \sum \alpha_i(m)m_i,$$
$$\text{where} \quad m_i, m \in M, \quad \alpha_i \in M^*.$$

For φ and ψ, the two required associativity relations $(*)$ from Sect. 2.1 hold. Consequently, we have a pre-equivalence situation $(R, S, M, M^*, \varphi, \psi)$ and the corresponding formal matrix ring. This ring has the following properties.

Lemma 3.8.4

(1) *The mapping φ (resp., ψ) is surjective if and only if M is an R-generator (resp., a finitely generated projective R-module).*

(2) *If M is an R-progenerator, then M satisfies the assumption and the assertions of Lemma 3.8.3.*

Proof (1). The image of the mapping φ is the sum of the images of all homomorphisms from M into R. It follows from the definition of a generator that φ is surjective if and only if M is a generator.

The mapping ψ is surjective if and only if the identity mapping of the module M is contained in the image of ψ, i.e., there exist homomorphisms $\alpha_1, \ldots, \alpha_k: M \to R$ and elements $m_1, \ldots, m_k$ such that $m = \sum \alpha_i(m)m_i$ for all $m \in M$. This means that $\{\alpha_1, \ldots, \alpha_k; m_1, \ldots, m_k\}$ is a dual basis of the module M. The last condition is equivalent to the property that M is a finitely generated projective R-module.

(2). The assertion directly follows from (1).   □

We formulate a result about equivalence categories, which is sometimes called the *first Morita theorem*.

Theorem 3.8.5 (The first Morita theorem) *Let the ring $K = \begin{pmatrix} R & M \\ N & S \end{pmatrix}$ be an equivalence situation. In such a case, the categories R-Mod, S-Mod, and K-Mod are equivalent to each other (the corresponding equivalences are given in the proof).*

Proof We define a functor $T_N = N \otimes_R (-) \colon R\text{-Mod} \to S\text{-Mod}$ using the relation $T_N(X) = N \otimes_R X$ for R-modules X. The functor T_N maps the R-module homomorphisms onto the induced S-module homomorphisms. The functor T_M is defined similarly. (Information about the functors T_N, T_M and some other functors is given in Sect. 3.1.)

We prove that the functors T_N and T_M are mutually inverse equivalences between the categories R-Mod and S-Mod. We have to verify that the composition $T_M T_N$ (resp., $T_N T_M$) is naturally equivalent to the identity functor of the category R-Mod (resp., S-Mod). This follows from the property that for any R-module X, there exist natural isomorphisms

$$(T_M T_N)X \cong (M \otimes_S N) \otimes_R X \cong R \otimes_R X \cong X.$$

Similarly, $(T_N T_M)Y \cong Y$ for an arbitrary S-module Y.

The functors

$$H_M = \mathrm{Hom}_R(M, -), \qquad H_N = \mathrm{Hom}_S(N, -), \qquad \text{where}$$
$$H_M(X) = \mathrm{Hom}_R(M, X), \quad H_N(Y) = \mathrm{Hom}_S(N, Y)$$

also define an equivalence of categories R-Mod and S-Mod, and homomorphisms are transformed into the induced homomorphisms. Indeed,

$$(H_N H_M)X = \mathrm{Hom}_S(N, \mathrm{Hom}_R(M, X)) \cong \mathrm{Hom}_R(M \otimes_S N, X)$$
$$\cong \mathrm{Hom}_R(R, X) \cong X, \qquad (H_M H_N)Y \cong Y.$$

We remark that the functors T_N and T_M, H_M and H_N are closely related to the functors T and H defined in Sect. 3.1. More precisely, the functors T_N and H_M are naturally equivalent. The same holds for the functors T_M and H_N. The homomorphism h is a natural isomorphism between $T_N(X)$ and $H_M(X)$, where

$$(h(n \otimes x))m = (mn)x, \qquad n \in N, \quad x \in X, \quad m \in M.$$

By Lemma 3.8.1, the homomorphism h is an isomorphism, since h is the homomorphism of module multiplication for the K-module $(X, H_M(X))$.

Now we take the functors $(1, T_N) \colon R\text{-Mod} \to K\text{-Mod}$ and $(1, 0) \colon K\text{-Mod} \to R\text{-mod}$ which are defined in Sect. 3.1. We have $((1, 0)(1, T_N))X = X$ and

$$((1, T_N)(1, 0))(X, Y) = (1, T_N)X = (X, T_N(X)) \cong (X, Y),$$

where the homomorphism of module multiplication g is taken as the isomorphism between $T_N(X)$ and Y; see Lemma 3.8.1 and Corollary 3.8.2. Thus, $(1, T_N)$ and $(1, 0)$ are mutually inverse equivalences of categories R-Mod and K-Mod. The functors $(1, H_M)$ and $(1, 0)$ are similarly defined; they play the same role $((1, H_M)$ is the restriction of the functor H from Sect. 3.1). The equivalence of categories S-Mod and K-Mod can be similarly proved.

Both equivalences can be obtained by applying the first part of the proof. For this purpose, we take a standard pre-equivalence situation arising from the use of the module $R \oplus M$. We also write this module in the form (R, M) to emphasize that we are dealing with an R-K-bimodule. Since $\text{End}_R(R \oplus M) \cong K$ and $\text{Hom}_R((R, M), R) \cong (R, N)$ $((R, N)$ is a K-R-bimodule), we have the corresponding matrix ring $\begin{pmatrix} R & (R, M) \\ \begin{pmatrix} R \\ N \end{pmatrix} & K \end{pmatrix}$; this is the ring K_1 from Sect. 2.3. In fact, we deal with an equivalence situation by Lemma 3.8.4. It follows from the above that the functors $T_{(R,N)}$ and $T_{(R,M)}$ define an equivalence of categories R-Mod and K-Mod. Substantially, $(1, T_N)$ and $(1, 0)$ are these functors. □

Under the conditions of Theorem 3.8.5, we say that an equivalence situation $(R, S, M, N, \varphi, \psi)$ or the corresponding matrix ring $\begin{pmatrix} R & M \\ N & S \end{pmatrix}$ defines the equivalence categories R-Mod and S-Mod.

The second Morita theorem asserts that the equivalence of two categories of modules appear from an equivalence situation.

Theorem 3.8.6 (The second Morita theorem) *Let R and S be two rings such that the categories R-Mod and S-Mod are equivalent. Then every equivalence of categories R-Mod and S-Mod is defined by some ring $\begin{pmatrix} R & M \\ N & S \end{pmatrix}$.*

Proof We assume that the functors $F \colon R\text{-Mod} \to S\text{-Mod}$ and $G \colon S\text{-Mod} \to R\text{-Mod}$ define mutually inverse equivalences. It is sufficient to prove that there exists an equivalence situation $(R, S, M, N, \varphi, \psi)$ or the ring $\begin{pmatrix} R & M \\ N & S \end{pmatrix}$. Indeed, it follows from the previous theorem that the functors T_N, T_M (and H_M, H_N) define an equivalence between R-Mod and S-Mod. However, it is known that, in this case, the functor F is equivalent to T_N and the functor G is equivalent to T_M.

We denote by M the R-module $G(S)$. We can turn it into in a right S-module such that M is an R-S-bimodule. This is done as follows. For two elements $m \in M$ and $s \in S$, we assume that $ms = \alpha(m)$, where the endomorphism α of the R-module M corresponds to s under the composition of ring isomorphisms $S \cong \text{End}_S S \cong \text{End}_R M$, where the second isomorphism is a well-known property of category equivalences. We assume that $S = \text{End}_R M$ and consider a standard pre-equivalence situation, defined by the bimodule M, and the corresponding ring $\begin{pmatrix} R & M \\ N & S \end{pmatrix}$, where $N = \text{Hom}_R(M, R)$. Under equivalences of categories, generators (resp., finitely generated projective modules) pass to modules with the same property (such properties are called *category* properties). Consequently, M is an R-progenerator. Thus, $\begin{pmatrix} R & M \\ N & S \end{pmatrix}$ is an equivalence situation by Lemma 3.8.4; which is what was required. □

Two rings R and S are said to be *equivalent* (in the sense of Morita) or *Morita-equivalent* if the categories R-Mod and S-Mod are equivalent. The notion of Morita-equivalence is left-right symmetrical. If R-Mod and S-Mod are equivalent, then it follows from Theorems 3.8.6 and 3.8.5 that the ring $\begin{pmatrix} R & M \\ N & S \end{pmatrix}$ is an equivalence situation and conversely. Then the opposite ring $\begin{pmatrix} R^\circ & N \\ M & S^\circ \end{pmatrix}$ (see Sect. 2.1) is an equivalence situation, and conversely. Consequently, the categories R°-Mod and S°-Mod are equivalent. Therefore, the categories Mod-R and Mod-S are equivalent.

Corollary 3.8.7 *For the rings R and S, the following conditions are equivalent.*

(1) *The rings R and S are equivalent.*
(2) *There exists an R-progenerator M such that $S \cong \operatorname{End}_R M$.*
(3) *There exists a right R-progenerator N such that $S \cong \operatorname{End}_R N$.*
(4) *There exists an equivalence situation $\begin{pmatrix} R & M \\ N & S \end{pmatrix}$.*

Proof The implications $(1) \Rightarrow (2)$ and $(1) \Rightarrow (3)$ follow from Theorem 3.8.6.

The equivalence $(1) \Leftrightarrow (4)$ is proved in Theorems 3.8.5 and 3.8.6.

$(2) \Rightarrow (1)$. We consider a standard pre-equivalence situation (R, S, M, M^*); see the remark before Lemma 3.8.4. The rings R and S are equivalent by Lemma 3.8.4 and Theorem 3.8.5.

$(3) \Rightarrow (2)$. The R°-module N is a progenerator and $S^\circ \cong \operatorname{End}_{R^\circ} N$. It follows from the above that the rings R° and S° are equivalent. Therefore, the rings R and S are also equivalent. $\square$

Corollary 3.8.8 *Let R be a ring and M an R-module.*

(1) *If M is a progenerator, then the rings R and $\operatorname{End}_R M$ are equivalent.*
(2) *For any positive integer n, the ring $M(n, R)$ of all $n \times n$ matrices is equivalent to the ring R.*

Proof (1). The assertion follows from Corollary 3.8.7.

(2). The assertion follows from the property that the ring $M(n, R)$ is isomorphic to the endomorphism ring of the free module R^n which is a progenerator. $\square$

We return to modules over the formal matrix ring $\begin{pmatrix} R & M \\ N & S \end{pmatrix}$. If the trace ideals I and J of this ring coincide with R and S, respectively (i.e., K is an equivalence situation), then all three categories R-Mod, S-Mod and K-Mod are equivalent to each other. Similarly, the categories of right modules Mod-R, Mod-S and Mod-K are also equivalent. We remark that the equivalence can be defined with the use of functors of the tensor products and Hom; their forms are given in the proof of Theorem 3.8.5. These functors preserve all module properties of category type. In particular, these functors preserve flat modules, projective modules and hereditary modules, which were considered earlier. Thus we have the following result.

Corollary 3.8.9 *Let* $K = \begin{pmatrix} R & M \\ N & S \end{pmatrix}$ *be an equivalence situation and* (A, B) *a K-module. Then the following conditions are equivalent.*

(1) *A is a flat (resp., projective, hereditary) R-module;*
(2) *B is a flat (resp., projective, hereditary) S-module;*
(3) *(A, B) is a flat (resp., projective, hereditary) K-module.*

Proof The assertion follows from Propositions 3.6.3 and 3.7.1 and the above K-module isomorphisms $(A, T(A)) \cong (A, B) \cong (T(B), B)$. $\qquad\square$

Corollary 3.8.10 *Under the conditions of Corollary 3.8.9, the following conditions are equivalent.*

(1) *The ring K is left (resp., right) hereditary.*
(2) *The ring R is left (resp., right) hereditary.*
(3) *The ring S is left (resp., right) hereditary.*

Corollary 3.8.11 *Let* $K = \begin{pmatrix} R & M \\ N & S \end{pmatrix}$ *be an equivalence situation.*

(1) *If (A, B) is a K-module, then the mappings $X \to NX$ and $X \to (X, NX)$ are isomorphisms from the lattice of all submodules of the module A onto the lattice of all submodules of the module B and the lattice of all submodules of the module (A, B), respectively. In the first case, the converse isomorphism is defined by the rule $Y \to MY$, where Y is an arbitrary submodule of B.*
(2) *The correspondence $L \to NL$ is an isomorphism from the lattice of all left ideals of the ring R onto the lattice of all submodules of the S-module N and ideals pass to subbimodules of the bimodule N. A similar assertion holds for the ring S and the bimodule M.*
(3) *The analogous assertions of (1) and (2) hold for right modules.*
(4) *The mappings $X \to NXM$ and $Y \to MYN$ are mutually inverse isomorphisms between the ideal lattices of rings R and S.*
(5) *The correspondence $X \to \begin{pmatrix} X & XM \\ NX & NXM \end{pmatrix}$ is an isomorphism from the lattice of ideals of the ring R onto the lattice of ideals of the ring K.*

Proof (1). The assertion follows from Corollary 3.8.2.
(2). The assertion follows from (1) applied to the K-modules (R, N) and (M, S).
(3). The assertion follows from (1), (2) and the symmetry argument.
(4). The assertion follows from (2).
(5). The assertion follows from (4) applied to the ring $\begin{pmatrix} R & (R, M) \\ \begin{pmatrix} R \\ N \end{pmatrix} & K \end{pmatrix}$, from the proof of Theorem 3.8.5. $\qquad\square$

In the next section, the following two well-known facts are used.

Corollary 3.8.12 *For a ring R, the following conditions are equivalent.*

(1) *R is a left hereditary ring.*
(2) *For any non-zero idempotent $e \in R$, the ring eRe is left hereditary.*
(3) *For any positive integers n, the ring $M(n, R)$ is left hereditary.*
(4) *There exists a positive integer n such that the ring $M(n, R)$ is left hereditary.*

A similar assertion holds for right hereditary rings.

Proof The implication $(1) \Rightarrow (2)$ follows from Proposition 3.7.5 if we consider the ring R as a formal matrix ring; see Sect. 2.1.

The implications $(2) \Rightarrow (1)$ and $(3) \Rightarrow (4)$ are obvious.

$(4) \Rightarrow (1)$. There exists a non-zero idempotent $e \in M(n, R)$ such that $R \cong e M$ $(n, R) e$. Since the implication $(1) \Rightarrow (2)$ is proved and the ring $M(n, R)$ is left (right) hereditary, the ring R is left (right) hereditary.

$(1) \Rightarrow (3)$. The ring $M(n, R)$ is left hereditary if and only if the module of column vectors $(R, \dots, R)$ of length n is a hereditary $M(n, R)$-module. This module can be considered as the module $(R, (R, \dots, R))$ over the formal matrix ring

$$\begin{pmatrix} R & (R, \dots, R) \\ \begin{pmatrix} R \\ \vdots \\ R \end{pmatrix} & M(n-1, R) \end{pmatrix}$$ of order 2. Then we apply Corollary 3.8.9 to the last module

and obtain that the ring $M(n, R)$ is left hereditary. $\square$

If K is an equivalence situation, then the study of K-modules can almost always be reduced to the study R-modules or S-modules. We consider another extreme case, where the trace ideals of the ring K are equal to the zero ideal. The study of K-modules can then often be reduced to the study of R-modules or S-modules, but additional difficulties appear. This is confirmed by our studies. The "intermediate" situation, in which I and J are non-trivial ideals, is quite complicated.

Corollary 3.8.9 completely describes flat, projective and hereditary modules provided K is an equivalence situation. On the other hand, Theorems 3.6.5, 3.7.3 and 3.7.6 contain satisfactory characterizations of such modules over a ring K with zero trace ideals. The structure of flat, projective, or hereditary modules over an arbitrary formal matrix ring K remain open questions. We can add regular modules to this list. (A module M is said to be *regular* if every cyclic submodule of M is a direct summand in M.) It is also important to determine, when the ring K is left hereditary or right hereditary.

We close this section with the following remark. Let M be some R-module. It is interesting to study the formal matrix ring $K = \begin{pmatrix} R & M \\ \mathrm{Hom}_R(M, R) & \mathrm{End}_R M \end{pmatrix}$ (a standard pre-equivalence situation) and two subrings of triangular matrices in K.

3.9 Hereditary Endomorphism Rings of Abelian Groups

In this section, we apply Corollary 3.7.9 to the description of some Abelian groups with hereditary endomorphism rings. The word "group" means an Abelian group. Groups are regarded as $\mathbb{Z}$-modules.

For a group G, we denote by $\operatorname{End} G$ the endomorphism ring of G. Let the group G be equal to the direct sum $A \oplus B$. Then we can identify the ring $\operatorname{End} G$ with the formal matrix ring $\begin{pmatrix} \operatorname{End} A & \operatorname{Hom}(A, B) \\ \operatorname{Hom}(B, A) & \operatorname{End} B \end{pmatrix}$. If B is a fully invariant subgroup, then $\operatorname{Hom}(B, A) = 0$ and we obtain a ring of triangular matrices. We often encounter this situation.

We use the following notations:

p is some prime integer;

$\mathbb{Z}(p)$ is a cyclic group of order p;

$\mathbb{Z}(p^\infty)$ is a quasicyclic p-group;

$\mathbb{Q}$ is the additive group or the field of rational numbers;

$\widehat{\mathbb{Z}}_p$ is the group or the ring of p-adic integers.

There are ring isomorphisms $\operatorname{End} \mathbb{Z}(p^\infty) \cong \widehat{\mathbb{Z}}_p$, $\operatorname{End} \mathbb{Q} \cong \mathbb{Q}$.

We often deal with divisible groups. A divisible group D can be represented in the form $D = \oplus_p D_p \oplus D_0$, where D_p is a divisible p-group, D_0 is a divisible torsion-free group and D_p is either the zero group or the direct sum of some set of copies of the group $\mathbb{Z}(p^\infty)$; D_0 is either the zero group or the direct sum of some set of copies of the group $\mathbb{Q}$.

Remark All used notions, facts and notations from the theory of Abelian groups are contained in the book [36].

A torsion group is said to be *elementary* if the order of each of its non-zero elements is not divisible by the square of an integer. Elementary groups are direct sums of elementary p-groups. Elementary non-zero p-groups are direct sums of the groups $\mathbb{Z}(p)$.

For a group G, the largest p-subgroup of G is called the *p-component* of the group G.

We often use Corollary 3.8.12. If $G = A \oplus B$ and the ring $\operatorname{End} G$ is left (resp., right) hereditary, then the rings $\operatorname{End} A$ and $\operatorname{End} B$ are left (resp., right) hereditary. Indeed, if e is the projection from the group G onto A with kernel B, then the ring $\operatorname{End} A$ can be identified with the ring $e \cdot \operatorname{End} G \cdot e$. In addition, we formulate the following result.

Proposition 3.9.1 ([73, Proposition 35.11])

(1) *If G is a group and ring $\operatorname{End} G$ is left or right hereditary, then G is not an infinite direct sum of non-zero groups.*

(2) *Let A be a reduced group with left or right hereditary ring $\operatorname{End} A$. Then every p-component A_p of the group A is an elementary p-group of finite rank and $A = A_p \oplus B_p$ for some group B_p.*

(3) *A reduced torsion group G has a left or right hereditary endomorphism ring if and only if G is an elementary group of finite rank (i.e., G is a finite direct sum of the groups $\mathbb{Z}(p)$ for some p).*

For convenience, we repeat Corollary 3.7.9 and formulate its right-side analogue.

Proposition 3.9.2 *Let $K = \begin{pmatrix} R & M \\ 0 & S \end{pmatrix}$.*

(1) *The ring K is left hereditary if and only if R and S are left hereditary rings, M is a flat S-module and M/ML is a projective R-module for any left ideal L of the ring S.*

(2) *The ring K is right hereditary if and only if R and S are right hereditary rings, M is a flat R-module and M/LM is a projective S-module for any right ideal L of the ring S.*

Remark Let D be some divisible group. It follows from Proposition 3.9.1 that when studying the groups with hereditary endomorphism rings, we can assume that D is a group of finite rank. We represent the group D in the form $D = D_0 \oplus D_t$, where D_0 is a torsion-free group and D_t is a torsion group. For $D_0 \neq 0$ and $D_t \neq 0$, the ring $\operatorname{End} D$ is the ring of triangular matrices $\begin{pmatrix} \operatorname{End} D_0 & \operatorname{Hom}(D_0, D_t) \\ 0 & \operatorname{End} D_t \end{pmatrix}$. For example, if $D = \mathbb{Q} \oplus \mathbb{Z}(p^\infty)$, then $\operatorname{End} D = \begin{pmatrix} \mathbb{Q} & A_p \\ 0 & \widehat{\mathbb{Z}}_p \end{pmatrix}$, where $A_p = \operatorname{Hom}(\mathbb{Q}, \mathbb{Z}(p^\infty))$ is the additive group of the field of p-adic numbers. This ring is not right hereditary, since A_p is not a projective $\widehat{\mathbb{Z}}_p$-module; see Proposition 3.9.2(2). At the same time, this ring is left hereditary, since all conditions of Proposition 3.9.2(2) hold. The same holds for the endomorphism ring of the group $\mathbb{Q} \oplus \mathbb{Z}(p_1^\infty) \oplus \ldots \oplus \mathbb{Z}(p_k^\infty)$, where $p_1, \ldots, p_k$ are distinct prime integers.

Below, we prove Theorem 3.9.3, which answers the following questions.

(1) When is the endomorphism ring of a divisible group right hereditary?
(2) When is the endomorphism ring of a divisible group left hereditary?

Theorem 3.9.3 *Let D be a non-zero divisible group of finite rank.*

(1) *The ring $\operatorname{End} D$ is right hereditary if and only if either D is a torsion-free group or D is a torsion group.*

(2) *The ring $\operatorname{End} D$ is left hereditary.*

Proof (1). Let the ring $\operatorname{End} D$ be right hereditary. By the above remark, the group D does not contain direct summands of the form $\mathbb{Z}(p^\infty) \oplus \mathbb{Q}$. Therefore, either D is a torsion-free group or D is a torsion group.

Conversely, if D is a torsion group, then $\operatorname{End} D$ is the finite direct product of matrix rings over rings of p-adic integers. If D is a torsion-free group, then $\operatorname{End} D$ is a matrix ring over $\mathbb{Q}$. In both cases, the ring $\operatorname{End} D$ is right and left hereditary by Corollary 3.8.12.

(2). If D is either a torsion group or a torsion-free group, then the ring End D is left hereditary; see the proof of (1). Let D be a mixed group, i.e., D contains a quasi-cyclic group and the group $\mathbb{Q}$. We denote by C the group $\mathbb{Q} \oplus \mathbb{Z}(p_1^\infty) \oplus \ldots \oplus \mathbb{Z}(p_k^\infty)$, where p_i are all prime integers such that the group D has a direct summand of the form $\mathbb{Z}(p_i^\infty)$. There exist a positive integer n and a group E such that $C^n \cong D \oplus E$. By Corollary 3.8.12 and the remark before the theorem, the ring End C^n is left hereditary. Therefore, the ring End D is left hereditary. $\qquad\square$

Now we begin to solve the following problem. We want to reduce the study of (right or left) heredity of the ring End G to the case where G is a reduced group.

Let G be a non-reduced non-divisible group. Then $G = A \oplus D$, where A is a non-zero reduced group and D is a non-zero divisible group. These notations are fixed until the end of this section. The ring End G coincides with the ring $\begin{pmatrix} \text{End } A & \text{Hom}(A, D) \\ 0 & \text{End } D \end{pmatrix}$.

Theorem 3.9.4 *The ring* End G *is right hereditary if and only if D is a torsion-free group of finite rank, the ring* End A *is right hereditary and* Hom(A, D) *is a flat* End A-*module.*

Proof Let the ring End G be right hereditary. It follows from Theorem 3.9.3 that the group D cannot be mixed. In addition, D is not a torsion group. Indeed, otherwise D is a finite direct sum of quasi-cyclic p-groups for some prime integer p. By Proposition 3.9.2, Hom(A, D) is a projective End D-module. (We remark that End D is a finite direct product of matrix rings over rings $\widehat{\mathbb{Z}}_p$.) The group structure of the group Hom(A, D) is known (see [36, Theorem 47.1]). Therefore, we obtain that Hom(A, D) cannot be a projective End D-module.

We assume now that D is a torsion-free group of finite rank, the ring End A is right hereditary and Hom(A, D) is a flat End A-module. Since End D is a matrix ring over $\mathbb{Q}$, it follows from Proposition 3.9.2 that the ring End G is right hereditary. $\quad\square$

We preserve the notations defined before Theorem 3.9.4. We pass to left hereditary rings.

Theorem 3.9.5 *The ring* End G *is left hereditary if and only if $G = T \oplus D$, where T is an elementary group of finite rank, D is a divisible group of finite rank and the groups T and D do not contain non-zero p-components for the same p.*

Proof Let the ring End G be left hereditary. It follows from Proposition 3.9.1 that the rank of the group D is finite. We assume that the group A contains elements of infinite order. By Proposition 3.9.2, Hom(A, D) is a projective End A-module. Now we observe that the additive group of the ring End A is reduced, since A is a reduced group. Therefore, Hom(A, D) is a reduced group. On the other hand, if D has a direct summand which is isomorphic to $\mathbb{Q}$, then Hom(A, D) also has a direct summand which is isomorphic to $\mathbb{Q}$. Similarly, if D contains the group $\mathbb{Z}(p^\infty)$, then Hom(A, D) is a non-reduced group. Thus, we obtain that A is a torsion group. Now it follows from Proposition 3.9.1 that the structure of the group A satisfies our Theorem.

The group $\mathbb{Z}(p) \oplus \mathbb{Z}(p^\infty)$ cannot be a direct summand of the group G. The reason is that the endomorphism ring of this group is the matrix ring $\begin{pmatrix} \mathbb{Z}/p\mathbb{Z} & \mathbb{Z}(p) \\ 0 & \widehat{\mathbb{Z}}_p \end{pmatrix}$. This ring is not left hereditary, since $\mathbb{Z}/p\mathbb{Z}$ is not a flat $\widehat{\mathbb{Z}}_p$-module. It follows from the above that the groups D and T have non-zero p-components only for distinct p.

Under the conditions of the theorem, we obtain that the subgroups D and T are fully invariant in G. Therefore, $\operatorname{End} G = \operatorname{End} D \times \operatorname{End} T$. The ring $\operatorname{End} D$ is left hereditary by Theorem 3.9.3, and the ring $\operatorname{End} T$ is left hereditary by Proposition 3.9.1.   □

Remarks Since $\operatorname{End}(\mathbb{Z} \oplus \mathbb{Q}) = \begin{pmatrix} \mathbb{Z} & \mathbb{Q} \\ 0 & \mathbb{Q} \end{pmatrix}$, it follows from Theorems 3.9.4 and 3.9.5 that this ring is right hereditary, but is not left hereditary.

Hereditary endomorphism rings of torsion-free groups are studied in the book [73]. In connection to Theorem 3.9.4, it is natural to raise the question of the description of groups A such that $\operatorname{Hom}(A, \mathbb{Q})$ is a flat $\operatorname{End} A$-module. For $\operatorname{End} A$-modules $\operatorname{Hom}(A, \mathbb{Z}(p))$ and $\operatorname{Hom}(A, \mathbb{Z}(p^\infty))$, it is interesting to determine when these modules are simple, Artinian, or Noetherian. A more detailed introduction to this field is contained in the book [73]; e.g., see Problems 11–13.

Chapter 4
Formal Matrix Rings over a Given Ring

If a concrete formal matrix ring of order n contains elements of some fixed ring R in all positions, then we speak about a *formal matrix ring over the ring* R. The properties of such rings can differ from those of the ordinary rings of all $n \times n$ matrices over R.

In Sects. 4.1 and 4.2, various properties of formal matrix rings over the ring R are given. There is a one-to-one correspondence between such rings and some sets of central elements of the ring R, which are called *multiplier systems*.

We formulate three problems about formal matrix rings over R. Sections 4.3–4.5 are related to the solutions of these problems.

In Sect. 4.6, we introduce the notion of the determinant of a formal matrix over a commutative ring. It is shown that this matrix invariant satisfies analogues of the main properties of the ordinary determinant. In addition, in Sect. 4.7, we show that some known theorems for ordinary determinants are also true for determinants of formal matrices.

4.1 Formal Matrix Rings over a Ring R

Let R be a ring. For any positive integer $n \geq 2$, there exists the matrix ring $M(n, R)$ of order n. We can define other matrix multiplications to obtain formal matrix rings. The corresponding bimodule homomorphisms $R \otimes_R R \to R$ for the ring $M(n, R)$ coincide and act using the rulation $x \otimes y \to xy$. By taking other bimodule homomorphisms $R \otimes_R R \to R$, we can obtain matrix rings over R of order n (as formal matrix rings) which are different from $M(n, R)$.

We take some ring K of formal matrices of order n, defined in Sect. 2.3, such that $R_1 = \ldots = R_n = R$ and $M_{ij} = R$ for all i and j. We call such a ring a *formal matrix ring of order n over the ring R*. Let $\varphi_{ijk} \colon R \otimes_R R \to R, i, j, k = 1, \ldots, n$, be the bimodule homomorphisms associated with the ring K. Similar to Sect. 2.3, if

© Springer International Publishing AG 2017
P. Krylov and A. Tuganbaev, *Formal Matrices*,
Algebra and Applications 23, DOI 10.1007/978-3-319-53907-2_4

$x, y \in R$, then we set $x \circ y = \varphi_{ijk}(x \otimes y)$. We also set $s_{ijk} = \varphi_{ijk}(1 \otimes 1)$ for every triple of subscripts i, j, k. For any elements $x, y \in R$ and subscripts i, j, k, we have

$$x \circ y = \varphi_{ijk}(x \otimes y) = x\varphi_{ijk}(1 \otimes 1)y = xs_{ijk}y.$$

Then we obtain that

$$xs_{ijk} = \varphi_{ijk}(x \otimes 1) = \varphi_{ijk}(1 \otimes x) = s_{ijk}x.$$

Thus, s_{ijk} is a central element of the ring R and $x \circ y = s_{ijk}xy$.

For all i, j, k, ℓ, the relations

$$s_{iik} = 1 = s_{ikk}, \quad s_{ijk} \cdot s_{ik\ell} = s_{ij\ell} \cdot s_{jk\ell} \tag{$*$}$$

hold. The first two relations follow from the property that φ_{iik} and φ_{ikk} coincide with the canonical isomorphism $R \otimes_R R \rightarrow R$, $x \otimes y \rightarrow xy$. The remaining relations follow from the property that $(x \circ y) \circ z = x \circ (y \circ z)$ for any $x, y, z \in R$; see the beginning of Sect. 2.3. In particular, the relation $(1 \circ 1) \circ 1 = 1 \circ (1 \circ 1)$ holds.

Now assume that we have an arbitrary set of central elements s_{ijk} of the ring R, $i, j, k = 1, \ldots, n$, which satisfy the relations $(*)$. For any three subscripts i, j, k, we can define a bimodule homomorphism

$$\varphi_{ijk}: R \otimes_R R \rightarrow R, \quad \varphi_{ijk}(x \otimes y) = s_{ijk}xy, \quad x, y \in R.$$

These homomorphisms define a formal matrix ring of order n in the sense of Sect. 2.3. Indeed, φ_{iik} and φ_{ikk} are canonical isomorphisms. Then we set $x \circ y = \varphi_{ijk}(x \otimes y)$ and obtain that $(x \circ y) \circ z = x \circ (y \circ z)$ for all $x, y, z \in R$ and corresponding subscripts i, j, k, ℓ.

Therefore, there is a one-to-one correspondence between formal matrix rings of order n over the ring R and sets $\{s_{ijk} \mid i, j, k = 1, \ldots, n\}$ of central elements of the ring R which satisfies the relations $(*)$. We denote such concrete rings by $M(n, R, \{s_{ijk}\})$ or $M(n, R, \Sigma)$, where $\Sigma = \{s_{ijk} \mid i, j, k = 1, \ldots, n\}$, or simply by the symbol K. The set Σ is called a *multiplier system* and its elements are called *multipliers* of the ring K. If all s_{ijk} are equal to 1, then we obtain the ring $M(n, R)$.

It is useful to give a formula for matrix multiplication in the ring $M(n, R, \Sigma)$. Namely, if $A = (a_{ij})$, $B = (b_{ij})$ and $AB = C = (c_{ij})$, then $c_{ij} = \sum_{k=1}^{n} s_{ikj}a_{ik}b_{kj}$.

We prove several identities for the multipliers s_{ijk} (we assume that we have a ring $M(n, R, \Sigma)$). We again write identities $(*)$ and call them the *main identities*:

$$s_{iik} = 1 = s_{ikk}, \quad s_{ijk} \cdot s_{ik\ell} = s_{ij\ell} \cdot s_{jk\ell}. \tag{4.1.1}$$

By setting $i = k$, we obtain $s_{iji} = s_{ij\ell} \cdot s_{ji\ell}$. Consequently, $s_{jij} = s_{ji\ell} \cdot s_{ij\ell}$. Therefore, the relation $s_{iji} = s_{jij}$ holds. For $j = \ell$, we have $s_{jkj} = s_{ijk} \cdot s_{ikj}$, whence $s_{iji} = s_{\ell ij} \cdot s_{\ell ji}$. Thus, we have

$$s_{iji} = s_{ij\ell} \cdot s_{ji\ell} = s_{\ell ij} \cdot s_{\ell ji}. \tag{4.1.2}$$

From identities (4.1.2), we obtain the following relations which follow from each other by a permutation of subscripts:

$$
\begin{aligned}
s_{iji} &= s_{jij} = s_{ijk} \cdot s_{jik} = s_{kij} \cdot s_{kji}, \\
s_{iki} &= s_{kik} = s_{ikj} \cdot s_{kij} = s_{jik} \cdot s_{jki}, \\
s_{jkj} &= s_{kjk} = s_{jki} \cdot s_{kji} = s_{ijk} \cdot s_{ikj}.
\end{aligned} \tag{4.1.3}
$$

The next two series of relations follow from identities (4.1.3):

$$
\begin{aligned}
s_{ijk} \cdot s_{iki} &= s_{jki} \cdot s_{iji} = s_{kij} \cdot s_{jkj}, \\
s_{kji} \cdot s_{iki} &= s_{ikj} \cdot s_{iji} = s_{jik} \cdot s_{jkj}.
\end{aligned} \tag{4.1.4}
$$

The relations (4.1.4) can be directly proved with the use of (4.1.1) if we substitute $\ell = i$ in (4.1.1) and interchange subscripts.

To a given ring $M(n, R, \Sigma)$, we can associate several matrices. Namely, we set

$$
S = (s_{iji}) = \begin{pmatrix} s_{111} & s_{121} & \cdots & s_{1n1} \\ s_{212} & s_{222} & \cdots & s_{2n2} \\ \cdots & \cdots & \cdots & \cdots \\ s_{n1n} & s_{n2n} & \cdots & s_{nnn} \end{pmatrix}.
$$

For every $k = 1, \ldots, n$, we form the matrix

$$
S_k = (s_{ikj}) = \begin{pmatrix} s_{1k1} & s_{1k2} & \cdots & s_{1kn} \\ s_{2k1} & s_{2k2} & \cdots & s_{2kn} \\ \cdots & \cdots & \cdots & \cdots \\ s_{nk1} & s_{nk2} & \cdots & s_{nkn} \end{pmatrix}.
$$

We call the matrices S, S_k *multiplier matrices* of the ring $M(n, R, \Sigma)$. The matrix S is symmetric. The main diagonal of the matrix S_k coincides with the kth row (and the kth column) of the matrix S, and the kth row and the kth column of the matrix S_k consist of 1s.

We formulate several interrelated problems about formal matrix rings $M(n, R, \Sigma)$.

(I). The realization and characterization problem. Assume that we have matrices $T, T_1, \ldots, T_n$ of order n with elements from the center $C(R)$. Under which conditions, are these matrices multiplier matrices of some ring $M(n, R, \Sigma)$? It is clear that the matrix T is necessarily symmetric. In addition, we can assume that each matrix T_k is also symmetric.

(II). The classification problem. Describe formal matrix rings depending on multiplier systems or multiplier matrices.

(III). The isomorphism problem. When do two multiplier systems define isomorphic formal matrix rings? In a more general setting, the isomorphism problem is formulated in Sect. 2.1.

Sections 4.3–4.5 are devoted to the above Problems (I)–(III). Now we consider some simple methods of constructing multiplier systems and a standard situation where that formal matrix rings are isomorphic.

(a). We can define the action of the symmetric group of degree n on multiplier systems and, consequently, on formal matrix rings. The corresponding orbits consist of isomorphic rings.

Let τ be a permutation of degree n. The action of τ on matrices is known. Namely, if $A = (a_{ij})$ is a matrix of order n, then we assume that $\tau A = (a_{\tau(i)\tau(j)})$. We mean that the matrix τA has the element $a_{\tau(i)\tau(j)}$ in position (i, j), and the element a_{ij} passes to position $(\tau^{-1}(i), \tau^{-1}(j))$.

Now, if $\Sigma = \{s_{ijk} \mid i, j, k = 1, \ldots, n\}$ is some multiplier system, then we set $t_{ijk} = s_{\tau(i)\tau(j)\tau(k)}$. Then $\{t_{ijk} \mid i, j, k = 1, \ldots, n\}$ is also a multiplier system, since it satisfies the relations (4.1.1). We denote it by $\tau\Sigma$. Consequently, there exists a formal matrix ring $M(n, R, \tau\Sigma)$. The rings $M(n, R, \Sigma)$ and $M(n, R, \tau\Sigma)$ are isomorphic under the correspondence $A \to \tau A$.

We also remark that the action of permutations on matrices can be represented in another form; see [59]. Let $T = (\delta_{i\tau(j)})$ be the matrix of a permutation τ, where $\delta_{i\tau(j)}$ is the Kronecker symbol. Then $\tau A = T^{-1}AT$, where $T^{-1} = (\delta_{i\tau^{-1}(j)})$. The sense of the isomorphism $M(n, R, \Sigma) \cong M(n, R, \tau\Sigma)$ can be expressed by the relation

$$T^{-1}(A \circ B)T = (T^{-1}AT) \circ (T^{-1}BT),$$

where the left product $\circ$ is calculated in $M(n, R, \Sigma)$ and the right product $\circ$ is calculated in $M(n, R, \tau\Sigma)$.

Let τ be a permutation such that $s_{ijk} = s_{\tau(i)\tau(j)\tau(k)}$ for all $i, j, k = 1, \ldots, n$. Then the correspondence $A \to \tau A$ is an automorphism of the ring $M(n, R, \Sigma)$. Let the ring R be commutative. We denote by Σ the multiplier system such that $s_{iik} = 1 = s_{ikk}$ and $s_{ijk} = 0$ provided $i \neq j$ and $j \neq k$. Then every permutation τ of degree n defines an automorphism of the R-algebra $M(n, R, \Sigma)$ which is not inner. More precisely, the automorphism group of this R-algebra is isomorphic to the semidirect product of the group of inner automorphisms by the symmetric group S_n.

(b). As above, Σ is a multiplier system s_{ijk} and α is an endomorphism of the ring R. Set $t_{ijk} = \alpha(s_{ijk})$. Then $\{t_{ijk}\}$ is a multiplier system. We denote it by $\alpha\Sigma$. There exists a ring homomorphism

$$M(n, R, \Sigma) \to M(n, R, \alpha\Sigma), \qquad (a_{ij}) \to (\alpha a_{ij}).$$

It is an isomorphism if α is an automorphism of the ring R.

(c). If $\Sigma = \{s_{ijk}\}$ and $X = \{x_{ijk}\}$ are two multiplier systems, then $\{x_{ijk}s_{ijk}\}$ is also a multiplier system. We denote it by $X\Sigma$. For every $\ell = 1, \ldots, n$, there exists a homomorphism

$$\zeta: M(n, R, X\Sigma) \to M(n, R, \Sigma), \qquad (a_{ij}) \to (x_{ij\ell}a_{ij}).$$

If all multipliers x_{ijk} are invertible elements, then ζ is an isomorphism.

Proof We take matrices $A = (a_{ij})$ and $B = (b_{ij})$ in the ring $M(n, R, X\Sigma)$. We have $AB = (c_{ij})$, where

$$c_{ij} = \sum_{k=1}^{n} x_{ikj} s_{ikj} a_{ik} b_{kj}.$$

The element

$$\sum_{k=1}^{n} x_{ij\ell} x_{ikj} s_{ikj} a_{ik} b_{kj}$$

stands in position (i, j) of the matrix $\zeta(AB)$. We have $\zeta(A) = (x_{ij\ell} a_{ij})$, $\zeta(B) = (x_{ij\ell} b_{ij})$, and the element

$$\sum_{k=1}^{n} s_{ikj} x_{ik\ell} x_{kj\ell} a_{ik} b_{kj}$$

stands in position (i, j) of the matrix $\zeta(A)\zeta(B)$. It follows from identities (4.1.1) that $x_{ikj} \cdot x_{ij\ell} = x_{ik\ell} \cdot x_{kj\ell}$. Therefore, ζ preserves products. If all x_{ijk} are invertible, then ζ has the inverse $\zeta^{-1} \colon (c_{ij}) \to (x_{ij\ell}^{-1} c_{ij})$. $\qquad\square$

Let E_{ij} denote matrix unit, i.e., the matrix E_{ij} has 1 in position (i, j) and E_{ij} has zeros in the remaining positions. We remark that $E_{ij} \cdot E_{jk} = s_{ijk} E_{ik}$.

(**d**). For each multiplier system $\Sigma = \{s_{ijk}\}$, the set $\Sigma^t = \{t_{ijk} \mid t_{ijk} = s_{kji}, i, j, k = 1, \ldots, n\}$ is also a multiplier system. Therefore, for the ring $M(n, R, \Sigma)$, the ring $M(n, R, \Sigma^t)$ exists; it is denoted by K^t. In addition, if $S, S_1, \ldots, S_n$ is a multiplier matrix of the ring K, then $S, S_1^t, \ldots, S_n^t$ is a multiplier matrix of the ring K^t. The ring $M(n, R, \Sigma^t)$ is isomorphic to the ring which is opposite to the ring $M(n, R, \Sigma)$ (the opposite ring is defined in Sect. 2.1).

If R is a commutative ring, then the transposition $A \to A^t$ is an anti-isomorphism of the rings K and K^t. Thus, the relation $(AB)^t = B^t A^t$ holds, where the right product is calculated in K^t. Therefore, we obtain the following.

Proposition 4.1.1 *Let R be a commutative ring. The relation $(AB)^t = B^t A^t$ holds for any matrices A and B if and only if $s_{ikj} = s_{jki}$ for all subscripts i, k, j (the matrices $S_1, \ldots, S_n$ are symmetric).*

Proof The remaining implication follows from the relations

$$(E_{ik} \cdot E_{kj})^t = (s_{ikj} E_{ij})^t = s_{ikj} E_{ji},$$
$$E_{kj}^t \cdot E_{ik}^t = E_{jk} E_{ki} = s_{jki} E_{ji}. \qquad\square$$

Corresponding to (c), the ring $M(n, R, \Sigma\Sigma')$ also exists. All its multiplier matrices are symmetric.

(e). Kronecker product of matrices. Let n and m be two arbitrary positive integers, R a commutative ring, $K = M(n, R, \Sigma)$ and $L = M(m, R, \Theta)$ two formal matrix rings over R with multiplier systems $\Sigma = \{s_{ijk} \mid i, j, k = 1, \ldots, n\}$ and $\Theta = \{t_{ijk} \mid i, j, k = 1, \ldots, m\}$, respectively. For matrices $A = (a_{ij}) \in K$ and $B = (b_{ij}) \in L$ the *Kronecker product* $A \otimes B$ of matrices A and B is defined as the block matrix

$$A \otimes B = \begin{pmatrix} a_{11}B & a_{12}B & \ldots & a_{1n}B \\ a_{21}B & a_{22}B & \ldots & a_{2n}B \\ \ldots & \ldots & \ldots & \ldots \\ a_{n1}B & a_{n2}B & \ldots & a_{nn}B \end{pmatrix}.$$

It is easy to specify an explicit form of elements of the matrix $A \otimes B$. Let k be a positive integer with $1 \leq k \leq nm$. There exists a unique pair of positive integers k_1, k_2 such that $1 \leq k_1 \leq n$, $1 \leq k_2 \leq m$, and $n(k_1 - 1) + k_2 = k$. The element $a_{i_1 j_1} \cdot b_{i_2 j_2}$ appears in position (i, j) of the matrix $A \otimes B$.

For every triple of integers i, j, k with $1 \leq i, j, k \leq nm$, we set

$$p_{ijk} = s_{i_1 j_1 k_1} \cdot t_{i_2 j_2 k_2}, \quad \Pi = \{p_{ijk} \mid i, j, k = 1, \ldots, nm\}.$$

Then Π is a multiplier system; i.e., relations (4.1.1) hold for p_{ijk}. Consequently, there exists a formal matrix ring $M = M(nm, R, \Pi)$. In addition, if $S, S_1, \ldots, S_n$ and $T, T_1, \ldots, T_m$ are multiplier matrices of the rings K and L, respectively, then $S \otimes T$, $S_i \otimes T_j$, $i = 1, \ldots, n$; $j = 1, \ldots, m$, are multiplier matrices of the ring M. Therefore, we can conditionally write $\Pi = \Sigma \otimes \Theta$.

The correspondence $\sum_{i=1}^{k} A_i \otimes_R B_i \to \sum_{i=1}^{k} A_i \otimes B_i$, $A_i \in K$, $B_i \in L$, defines an isomorphism of R-algebras $K \otimes_R L$ and $M(nm, R, \Pi)$. There also exists an isomorphism $(K \otimes_R L)^o \cong K^o \otimes_R L^o$, which can be proved with the use of d).

The properties of Kronecker products of formal matrices are similar to those of Kronecker products of ordinary matrices.

Here are three such properties.

(1). If $A, C \in K$ and $B, D \in L$, then $(A \otimes B)(C \otimes D) = AC \otimes BD$.

(2). If A and B are invertible matrices, then $(A \otimes B)^{-1} = A^{-1} \otimes B^{-1}$.

(3). $d(A \otimes B) = d(A)^n \cdot d(B)^m$, where $d(X)$ is the determinant of the formal matrix X; this determinant will be defined in Sect. 4.6.

We consider a homomorphism which will be quite useful later. It is a particular case of the homomorphism ζ from (c).

Assume that we have a formal matrix ring $M(n, R, \Sigma)$, $\Sigma = \{s_{ijk}\}$. We fix the subscript $\ell = 1, \ldots, n$ and set $t_{ij} = s_{ij\ell}$ for all $i, j = 1, \ldots, n$. We remark that $t_{ii} = 1$ and the relation $s_{ijk} \cdot t_{ik} = t_{ij} \cdot t_{jk}$ follows from the main identities (4.1.1). In particular, $s_{iji} = t_{ij} \cdot t_{ji}$.

We define the mapping

$$\eta: M(n, R, \Sigma) \to M(n, R), \qquad (a_{ij}) \to (t_{ij}a_{ij}).$$

Proposition 4.1.2

(1) η is a ring homomorphism.

(2) If the multiplier s_{ikj} is divided by t_{ik} or t_{kj} for all i, j, k, then $\mathrm{Ker}(\eta)$ is a nilpotent ideal of nilpotence index 2.

(3) The mapping η is injective if and only if all s_{ijk} are non-zero-divisors.

(4) η is an isomorphism if and only if all elements s_{ijk} are invertible.

Proof **(1)**. The assertion follows from (c) if we take Σ to be the system consisting of 1s and set X equal to Σ.

(2). We take arbitrary matrices $A = (a_{ij})$, $B = (b_{ij})$ in the ring $M(n, R, \Sigma)$. If $\eta A = 0 = \eta B$, then $t_{ij}a_{ij} = 0 = t_{ij}b_{ij}$ for all i and j. It follows from the proof of (c) that $AB = 0$. □

To close this section, we turn our attention to formal matrix rings over the ring R of order 2 and 3. In the case $n = 2$, it is easy to achieve full clarity. Indeed, we take an arbitrary ring $M(2, R, \Sigma)$ and find out how the multiplication is performed in this ring.

From the previous material, it is easy to obtain the following property. There exists an element $s \in C(R)$ such that matrices in $M(2, R, \Sigma)$ are multiplied using the relation

$$\begin{pmatrix} a_{11} & a_{12} \\ a_{21} & a_{22} \end{pmatrix} \begin{pmatrix} b_{11} & b_{12} \\ b_{21} & b_{22} \end{pmatrix} = \begin{pmatrix} a_{11}b_{11} + sa_{12}b_{21} & a_{11}b_{12} + a_{12}b_{22} \\ a_{21}b_{11} + a_{22}b_{21} & sa_{21}b_{12} + a_{22}b_{22} \end{pmatrix}. \qquad (*)$$

We also have the converse. Every central element s of the ring R determines a ring $M(2, R, s)$ of formal 2×2 matrices over R whose multiplication satisfies the rule $(*)$. Such rings are defined in [70]; they are denoted by K_s in [70]. They are also studied and used in [43, 44, 61, 69, 74, 104, 105].

The classification of rings $M(3, R, \Sigma)$ is more complicated. The main identities (4.1.1) turn into identities

$$s_{iik} = 1 = s_{ikk}, \qquad i, k = 1, 2, 3,$$

$$s_{121} = s_{212} = s_{123} \cdot s_{213} = s_{321} \cdot s_{312},$$

$$s_{131} = s_{313} = s_{132} \cdot s_{312} = s_{231} \cdot s_{213},$$

$$s_{232} = s_{323} = s_{231} \cdot s_{321} = s_{132} \cdot s_{123}.$$

Every family of elements $\{s_{ijk} \mid i, j, k = 1, 2, 3\}$ satisfying the previous relations defines a formal matrix ring over R of order 3. If $s_{ikj} = s_{jki}$ for all $i, j, k = 1, 2, 3$, i.e., the matrices S_1, S_2, and S_3 are symmetric, then these relations can be represented more compactly:

$$s_{iik} = 1 = s_{ikk}, \qquad i, k = 1, 2, 3,$$
$$s_{121} = s_{212} = s_{123} \cdot s_{213},$$
$$s_{131} = s_{313} = s_{132} \cdot s_{312},$$
$$s_{232} = s_{323} = s_{231} \cdot s_{321}.$$

In connection to the problem (I), we can definitely say that not every symmetric matrix of order 3 can be the matrix S for some ring $M(3, R, \Sigma)$.

We determine the structure of K-modules for the ring K of formal matrices of order 2, i.e.

$$K = M(2, R, s) = \begin{pmatrix} R & R \\ R & R \end{pmatrix}.$$

We pay special attention to the modules $(A, T(A))$, $(T(B), B)$ and $(T(B), T(A))$. This is interesting in itself and will be used at the end of Sect. 5.5.

For any elements $x, y \in R$, we assume that $x \circ y = sxy$. Thus, $x \circ y$ coincides with the element which, in the case of an arbitrary ring K, is denoted by mn or nm.

Now assume that we have an arbitrary K-module (A, B) with homomorphisms of module multiplication

$$g: R \otimes_R A \to B, \qquad f: R \otimes_R B \to A.$$

We identify $R \otimes_R A$ and $R \otimes_R B$ with A and B, respectively, by assuming $r \otimes a = ra$ and $r \otimes b = rb$. Having done this, we can assume that we have homomorphisms $g: A \to B$ and $f: B \to A$. We introduce some notations. We define $y \circ a = g(ya)$ and $y \circ b = f(yb)$, where $y \in R$, $a \in A$ and $b \in B$.

For the ring $M(2, R, s)$, the relations $m(na) = (mn)a$ and $n(mb) = (nm)b$ turn into $x \circ (y \circ a) = (x \circ y)a$ and $y \circ (x \circ b) = (y \circ x)b$, where $x, y \in R$, $a \in A$, $b \in B$. Now we have

$$x \circ (y \circ a) = x \circ g(ya) = x \circ (yg(a)) = xf(yg(a)) = xyfg(a),$$
$$(x \circ y)a = sxya.$$

Therefore, $xyfg(a) = sxya$ for all $x, y \in R$. Consequently, $fg(a) = sa$, $a \in A$, and $fg = s \cdot 1_A$. Similarly, we have $gf = s \cdot 1_B$.

We also have the converse. Each pair of R-homomorphisms $g: A \to B$ and $f: B \to A$ which satisfy the relations $fg = s \cdot 1_A$ and $gf = s \cdot 1_B$ defines the K-module (A, B) with module multiplication

$$\begin{pmatrix} x & v \\ w & y \end{pmatrix} \begin{pmatrix} a \\ b \end{pmatrix} = \begin{pmatrix} xa + vf(b) \\ wg(a) + yb \end{pmatrix}.$$

Now let (A, B) be a K-module and $g: A \to B$, $f: B \to A$ the homomorphisms of module multiplication of (A, B). We proved above that they satisfy the relations

$fg = s \cdot 1_A$ and $gf = s \cdot 1_B$. We describe the construction of the K-module $(T(B), T(A))$. For this module, the homomorphisms of module multiplication are $f' = 1 \otimes f$ and $g' = 1 \otimes g$. (We recall that $T(B)$ is identified with B, $T(A)$ is identified with A, and so on.) The mapping $f' \colon R \otimes_R (R \otimes_R B) \to R \otimes_R A$ acts with the use of the relation $x \otimes (y \otimes b) = x \otimes y \circ b$. However, it follows from our agreements that

$$x \otimes (y \otimes b) = xyb \quad \text{and} \quad x \otimes y \circ b = x(y \circ b) = xyf(b).$$

Therefore, we have

$$f'(xyb) = xyf(b), \quad f'(b) = f(b), \quad f' = f.$$

Similarly, we have $g' = g$. We conclude that the K-module $(T(B), T(A))$ coincides with the K-module (B, A) with homomorphisms of module multiplication f and g.

There exist two homomorphisms

$$(g, f) \colon (A, B) \to (B, A) \quad \text{and} \quad (f, g) \colon (B, A) \to (A, B),$$

since $g(y \circ b) = y \circ f(b)$ and $f(x \circ a) = x \circ g(a)$ for all values of a, b, x, y.

For an arbitrary R-module A, we find the structure of the K-modules $(A, T(A))$ and $(T(A), A)$. Since we identify $T(A)$ with A, we denote the first module by $(A, A)'$, and we denote the second module by $(A, A)''$. For $(A, A)'$, the homomorphisms of module multiplication are 1_A and the mapping $x \otimes (y \otimes a) \to (x \circ y)a$. However, $x \otimes (y \otimes a) = xya$ and $(x \circ y)a = sxya$. Therefore, the second homomorphism of module multiplication is equal to $s \cdot 1_A$. Conversely, for the K-module $(A, A)''$, the homomorphisms of module multiplication are $s \cdot 1_A$ and 1_A.

Thus, each R-module A defines two K-modules: $(A, A)'$ and $(A, A)''$. The structure of the module $(T(A), T(A))$ is easily described; for the module $(A, A)'$, the module $(T(A), T(A))$ is $(A, A)''$, and conversely.

4.2 Some Properties of Formal Matrix Rings over R

We remark that Tang and Zhou [106] proved many of the results of this section for a class of formal matrix rings over R; see Sect. 4.3 for more about such rings.

Let $M(n, R, \Sigma)$ be a formal matrix ring of order n over a ring R. We recall that $\Sigma = \{s_{ijk} \mid i, j, k = 1, \ldots, n\}$ is a multiplier system. The multipliers s_{ijk} satisfy the relations (4.1.1) and their corollaries (4.1.3) and (4.1.4) from Sect. 4.1. Sometimes, we denote by R_{ij} the ring R in position (i, j).

We apply Theorems 2.4.3 and 2.4.4 to calculate the Jacobson radical and the prime radical of the ring $M(n, R, \Sigma)$. Using the notation of Theorem 2.4.3, we have $J(R_{ij}) = \{x \in R_{ij} \mid R_{ji} \circ x \subseteq J(R)\}$. Then $R_{ji} \circ x = s_{jij}Rx = s_{iji}Rx$. Therefore, $J(R_{ij}) = \{x \in R_{ij} \mid s_{iji}x \in J(R)\}$. We remark that the ideal $J(R_{ij})$ coincides with the intersection of all maximal left (right) ideals of the ring R which do not contain s_{iji}. We denote this ideal $J(R_{ij})$ by $J_{ij}(R)$. Now the following result follows from Theorem 2.4.3.

Corollary 4.2.1 *The Jacobson radical of the ring $M(n, R, \Sigma)$ is equal to*

$$
\begin{pmatrix}
J(R) & J_{12}(R) & \ldots & J_{1n}(R) \\
J_{21}(R) & J(R) & \ldots & J_{2n}(R) \\
\ldots & \ldots & \ldots & \ldots \\
J_{n1}(R) & J_{n2}(R) & \ldots & J(R)
\end{pmatrix}.
\tag{$*$}
$$

The prime radical of the ring $M(n, R, \Sigma)$ has a similar structure. Set $P(R_{ij}) = \{x \in R_{ij} \mid s_{iji} x \in P(R)\}$.

Corollary 4.2.2 *The prime radical of the ring $M(n, R, \Sigma)$ coincides with the ideal of matrices of the form $(*)$, only the symbol J needs to be replaced by P.*

Next we give an inner description of ideals of the ring $M(n, R, \Sigma)$; we also find the structure of the factor rings of this ring. For arbitrary formal matrix rings, these questions were briefly considered in Sects. 2.1 and 2.4. The following proposition is proved by direct calculation.

Proposition 4.2.3 *If I is an ideal of the ring $M(n, R, \Sigma)$, then*

$$
I =
\begin{pmatrix}
I_{11} & I_{12} & \ldots & I_{1n} \\
I_{21} & I_{22} & \ldots & I_{2n} \\
\ldots & \ldots & \ldots & \ldots \\
I_{n1} & I_{n2} & \ldots & I_{nn}
\end{pmatrix},
$$

where all I_{ij} are ideals in R and the relations

$$
I_{ii} \subseteq \cap_{\ell=1}^{n} (I_{i\ell} \cap I_{\ell i}), \quad s_{iji} I_{ij} \subseteq I_{ii} \cap I_{jj}
$$

hold for all $i, j = 1, \ldots, n$, and $s_{ikj} I_{kj} \subseteq I_{ij}$, $s_{jki} I_{jk} \subseteq I_{ji}$ for all distinct i, j, k.

The following proposition is also proved by standard arguments.

Proposition 4.2.4 *Let $I = (I_{ij})$ be an ideal of the ring $K = M(n, R, \Sigma)$.*

(1) *The set of matrices*

$$
\overline{K} =
\begin{pmatrix}
R/I_{11} & R/I_{12} & \ldots & R/I_{1n} \\
R/I_{21} & R/I_{22} & \ldots & R/I_{2n} \\
\ldots & \ldots & \ldots & \ldots \\
R/I_{n1} & R/I_{n2} & \ldots & R/I_{nn}
\end{pmatrix}
$$

is a formal matrix ring with bimodule homomorphisms

$$\varphi_{ijk} : R/I_{ij} \otimes_{R/I_{jj}} R/I_{jk} \to R/I_{ik},$$
$$\varphi_{ijk}(\overline{x} \otimes \overline{y}) = s_{ijk}xy + I_{ik}$$

for all $i, j, k = 1, \dots, n$.

(2) *There is an isomorphism*

$$K/I \cong \overline{K}, \qquad (x_{ij}) + I \to (x_{ij} + I_{ij}).$$

We consider several common ring properties in the case of our matrix rings. We recall that E_{ij} is a matrix unit.

Proposition 4.2.5 *Let K be a formal matrix ring of order n over the ring R with multipliers s_{ijk}, $i, j, k = 1, \dots, n$. The following assertions hold.*

(1) *K is a simple ring if and only if R is a simple ring and all multipliers s_{ijk} are not equal to zero.*

(2) *K is a prime ring if and only if R is a prime ring and all s_{ijk} are not equal to zero.*

(3) *K is a regular ring if and only if R is a regular ring and all elements s_{ijk} are invertible in R.*

(4) *K is a semiprimitive ring if and only if R is a semiprimitive ring and all s_{ijk} are not zero-divisors in R.*

(5) *K is a semiprime ring if and only if R is a semiprime ring and all s_{ijk} are not zero-divisors in R.*

Proof (1). $\Rightarrow$. If I is a non-zero ideal in R, then $\begin{pmatrix} I & \dots & I \\ & \dots\dots\dots & \\ I & \dots & I \end{pmatrix}$ is a non-zero ideal in K. Therefore, $I = R$, since the ring K is simple. Consequently, R is a simple ring. The homomorphism η from Proposition 4.1.2 has to be injective. Therefore, all s_{ijk} are non-zero.

$\Leftarrow$. The ring $M(n, R)$ is simple, and all central elements s_{ijk} are invertible, since the center of a simple ring is a field. By Proposition 4.1.2, $K \cong M(n, R)$; consequently, K is a simple ring.

(2). $\Rightarrow$. Since $R \cong E_{11}KE_{11}$, we have that R is a prime ring. We assume that $s_{ijk} = 0$ for some i, j, k. Then $E_{ij}KE_{k\ell} = 0$. This contradicts the property that K is a prime ring.

$\Leftarrow$. The center of a prime ring is a domain. Therefore, all s_{ijk} are non-zero-divisors. We assume that $(a_{ij})K(b_{ij}) = 0$ for some non-zero matrices (a_{ij}) and (b_{ij}). For example, let $a_{k\ell} \neq 0$ and $b_{mp} \neq 0$. Then $s_{k\ell m}s_{kmp}a_{k\ell}Rb_{mp} = 0$, whence $a_{k\ell}Rb_{mp} = 0$; this contradicts the property that R is a prime ring. Thus, K is a prime ring.

(3). $\Rightarrow$. Since the ring K is regular and $R \cong E_{11}KE_{11}$, the ring R is regular. Consequently, there exists a matrix (a_{ij}) such that $E_{ij} = E_{ij}(a_{ij})E_{ij}$. Therefore, we obtain the relation $s_{iji}a_{ji} = 1$, and all elements s_{iji} are invertible. It follows from relations (4.1.3) of Sect. 4.1 that all elements s_{ijk} are also invertible.

$\Leftarrow$. The ring $M(n, R)$ is regular and $K \cong M(n, R)$ by Proposition 4.1.2.

(4) and (5). The assertions follow from relations (4.1.3) of Sect. 4.1 and Corollaries 4.2.1 and 4.2.2, respectively. $\square$

4.3 Characterization of Multiplier Matrices

In this section and the following two sections R denotes an arbitrary ring, $M(n, R, \Sigma)$ is a formal matrix ring of order n over the ring R and Σ is a multiplier system $\{s_{ijk} \mid i, j, k = 1, \ldots, n\}$. The multipliers s_{ijk} satisfy identities (4.1.1) of Sect. 4.1. The element in position (i, j) of the product of matrices (a_{ij}) and (b_{ij}) from the ring $M(n, R, \Sigma)$ is equal to $\sum_{k=1}^{n} s_{ikj} a_{ik} b_{kj}$.

In Sect. 4.1, we formulated Problem (I) concerning matrices which are multiplier matrices for formal matrix rings. Problem (II) is related to the description of rings $M(n, R, \Sigma)$ depending on multiplier systems Σ. In general, it is difficult to solve these problems, since it is difficult to verify identities (4.1.1) of Sect. 4.1.

It seems that the situation is simpler if all multipliers s_{ijk} are integral powers of some central element s of the ring R. We know that this is always the case for $n = 2$; see the end of Sect. 4.1.

For $n \geq 3$, the characterization and classification of the rings $M(n, R, \Sigma)$ becomes more complicated, even if all multipliers s_{ijk} are powers of the element s. The special case is interesting, where every multiplier s_{ijk} is equal to s^m for some $m \geq 1$. Tang and Zhou [106] give a detailed study of rings $M(an, R, \Sigma)$ such that $s_{iji} = s^2$ provided $i \neq j$ and $s_{ijk} = s$ for all pairwise distinct i, j, k. We call such rings *Tang–Zhou rings*.

Here and in Sect. 4.4, we consider the case where $s_{ijk} = 1$ or $s_{ijk} = s$ for all i, j, k, where s is some central element of the ring R. We impose simple additional restrictions on the element s. Every corresponding matrix ring is denoted by $M(n, R, s)$; it is clear that for fixed n, there exists only a finite number of distinct such rings. The element s is called a *multiplier* of the ring $M(n, R, s)$.

We assume that there is a ring $M(n, R, s)$. In addition, we assume that $s^2 \neq 1$ and s is not an idempotent. Without great loss of generality, we may assume that the element s is not invertible. Indeed, if the element s is invertible, then $M(n, R, s) \cong M(n, R)$ by Proposition 4.1.2. The following condition is stronger: $s^k \neq s^\ell$ for any nonnegative distinct integers k and ℓ.

We assume that there is a ring $M(n, R, s)$, where the element s^2 is not equal to 1 or s; in particular, $s \neq 0$ and $s \neq 1$. We point out some relations for multipliers of the form s_{iji}.

Lemma 4.3.1 *Let the subscripts i, j, k be pairwise distinct. For the elements s_{iji}, s_{iki} and s_{jkj}, only one of the following three statements holds.*

(1) *All three of these elements are equal to 1.*
(2) *Two of the tree elements are equal to s, and the third element is equal to 1.*
(3) *All three elements are equal to s.*

Proof It directly follows from two of the relations (4.1.3) of Sect. 4.1 that the case where two of three multipliers s_{iji}, s_{iki}, s_{jkj} are equal to 1 and the third multiplier is equal to s is impossible. Therefore, we have only the possibilities mentioned in (4.1.1)–(4.1.3). All three cases are possible, as we will show later.   □

In Sect. 4.1, we constructed square matrices S, $S_1, \ldots, S_n$ of order n consisting of multipliers s_{ijk} of some formal matrix ring. We called these matrices *multiplier matrices* of the given ring. In what follows, the term "multiplier matrix" usually mean "the matrix S".

It is useful to define the notion of an abstract multiplier matrix. As earlier, let s be some central element of the ring R such that $s^2 \neq 1$ and $s^2 \neq s$. Let $T = (t_{ij})$ be a symmetric matrix of order n such that all its elements are equal to 1 or s, the main diagonal consists of 1s, and for any three elements t_{ij}, t_{ik}, t_{jk}, one of the assertions (1), (2), (3) of Lemma 4.3.1 holds. We call such a matrix T a *multiplier matrix*. If τ is some permutation of degree n, then $\tau T = (t_{\tau(i)\tau(j)})$ is also a multiplier matrix (matrices of the form τT are defined in Item a) of Sect. 4.1).

Let T be a matrix which can be represented in block form such that the blocks consisting of 1 s appear on the main diagonal and the element s appears in all remaining positions. It is clear that T is a multiplier matrix. In such a case, we say that T *is of the canonical form*.

Lemma 4.3.2 *For each multiplier matrix T, there exists a permutation σ such that the matrix σT is of the canonical form.*

Proof We use induction on n. For $n = 2$, there exists a unique multiplier matrix $\begin{pmatrix} 1 & s \\ s & 1 \end{pmatrix}$ which is of the canonical form.

We assume that the assertion of the lemma holds for all matrices of order $\leq n - 1$, where $n \geq 3$. Let T be a matrix of order n. We take the submatrix T' of order $n - 1$ standing in the right lower corner of the matrix T. There exists a permutation τ' of integers $2, \ldots, n$ such that the matrix $\tau' T'$ is of the canonical form. We take a permutation τ of degree n coinciding with τ' on integers $2, \ldots, n$. The matrix $\tau' T'$ appears in the right lower corner of the matrix τT. If all elements of the first rows of the matrix τT, beginning with the second element, are equal to s, then τT is already of the canonical form.

Now we assume that the first row of the matrix τT, excluding the position $(1,1)$, contains 1. Let the second block, standing on the main diagonal after position $(1,1)$, have order $m - 1$, $m \geq 2$. In this case, the elements $t_{12}, \ldots, t_{1m}$ are simultaneously equal to either 1 or s. Indeed, if $t_{1i} = 1$, $t_{1j} = s$, $2 \leq i, j, \leq m$, then $t_{ij} = s$, which is impossible. We assume that $t_{12} = \ldots = t_{1m} = 1$. If $t_{1\ell} = 1$ for some ℓ, $m + 1 \leq \ell \leq n$, then $t_{2\ell} = 1$, which is also impossible. Consequently, $t_{1\,m+1} = \ldots = t_{1n} = s$. Now, if we unite the first block and the second block from the main diagonal, then we obtain the canonical form of the matrix τT.

Thus, we can assume that $t_{12} = \ldots = t_{1m} = s$. What can we say about the elements $t_{1\,m+1} = \ldots = t_{1n}$? We assume that $t_{1i} = 1$, $t_{1j} = 1$, and $t_{1k} = 1$, where $m + 1 \leq i, j, k \leq n$ and $i < j < k$. In this case, $t_{ij} = s$, $t_{ik} = 1$; this is a contradiction. Therefore, the sequence $t_{1\,m+1}, \ldots, t_{1n}$ can only be of one of the following three types:

$$(1)\ 111\ldots sss; \quad (2)\ sss\ldots 111\ldots sss; \quad (3)\ sss\ldots 111.$$

It is important that, under the sequence $11\ldots 1$ from (1), (2) or (3), the main diagonal contains a block whose order is equal to the number of units in this sequence.

In cases (1) and (2), we take the submatrix whose right lower corner ends with the block corresponding to the sequence $11\ldots 1$. By some permutation, we reduce this submatrix to the canonical form. Then we add missing one-term cycles to the same permutation, apply the permutation to the whole matrix, and reduce it to the canonical form (under these actions, the submatrix standing in the right lower corner does not change).

We consider the remaining case (3) and the block on the main diagonal standing under the sequence $11\ldots 1$. Let the block begin with the row $k + 1$, where $k \geq 2$. We apply the cycle $(1\,2\ldots k)$ to the matrix. As a result, in the right lower corner, the block of order $k + 1$, consisting of 1s, will appear. In the first right row and in the first left column, s will stand instead of 1. The element s will also stand in the positions $(k, 1), \ldots, (k, k - 1)$, and $(1, k), \ldots, (k - 1, k)$. Thus, s stands in all positions in the right lower corner to the left and up the blocks consisting of 1. By some permutation, we reduce to the canonical form the submatrix appearing in rows and columns $1, 2, \ldots, k - 1$. Then the whole matrix will have the canonical form.   □

We present a summary of yet another proof of Lemma 4.3.2. The relation $\sim$, where $i \sim j \Leftrightarrow t_{ij} = 1$, is an equivalence relation on the set $\{1, 2, \ldots, n\}$. Let σ be a permutation whose upper row contains integers $1, \ldots, n$ in the natural order. The lower row consists of the equivalence classes with respect to relation $\sim$, the classes are arranged in an arbitrary order. Then the matrix σT is of the canonical form.   □

Under the conditions of the lemma, we say that the *matrix T is reduced to the canonical form σT*. We also have the converse; if some matrix T is reduced to the canonical form σT, then T is a multiplier matrix. Thus, multiplier matrices coincide with matrices which can be reduced to the canonical form by permutations.

Let a matrix T be of the canonical form. Then the blocks standing on the main diagonal of the matrix can be arranged in any preassigned order. In other words, there exists a permutation τ such that the blocks in the matrix τT are arranged in the required order, and τT is of the canonical form. For this purpose, it is sufficient to show that if T consists of two blocks, then we can interchange them. Let the first block have order k, $1 \leq k \leq n - 1$. Then we can take the permutation

$$\begin{pmatrix} 1 & 2 & \ldots n - k\ n - k + 1 \ldots n \\ k + 1\ k + 2 \ldots & n & 1 & \ldots k \end{pmatrix}$$

as τ.

We formulate the main result of this section. First, we remark that a special case appearing in the proof will be considered in the next section.

Theorem 4.3.3 *Let T be a multiplier matrix. There exists a ring $M(n, R, s)$, a multiplier matrix of which coincides with T.*

Proof We use induction on n. For $n = 2$, there exists only one multiplier matrix; namely, the matrix $\begin{pmatrix} 1 & s \\ s & 1 \end{pmatrix}$. We know that a ring $M(2, R, s)$ with multiplier matrix $\begin{pmatrix} 1 & s \\ s & 1 \end{pmatrix}$ always exists.

Now we assume that $n > 2$. Let σ be a permutation such that the matrix σT is of the canonical form (see Lemma 4.3.2). The following two cases are possible.

Case 1. All blocks from the main diagonal are of order 1. In Sect. 4.4, it will be shown that there exists a ring $M(n, R, s)$ with multiplier matrix σT.

Case 2. Not all blocks mentioned in case 1 are of order 1. Before the theorem, we showed that there exists a permutation τ such that the matrix $\tau \sigma T$ is of the canonical form, and the order of the lowest blocks exceeds 1. In $\tau \sigma T$, we take the submatrix T' of order $n - 1$ which is contained in the left upper corner. It is of the canonical form. By the induction hypothesis, there exists a ring K of the form $M(n - 1, R, s)$ with multiplier matrix T'. We represent this ring as a ring of block matrices of order 2:

$$\begin{pmatrix} R & \ldots & R & \vline & R \\ \cdot & \ldots & \cdot & \vline & \cdot \\ R & \ldots & R & \vline & R \\ \hline R & \ldots & R & \vline & R \end{pmatrix}.$$

Now we apply the construction of the ring K_4 from Sect. 2.3 to the ring K. The multiplier matrix of the obtained ring K_4 coincides with the matrix $\tau \sigma T$. The ring $\sigma^{-1} \tau^{-1} K_4$ is the required formal matrix ring. Indeed, the matrix T is the multiplier matrix of this ring.   □

We return to the action of permutations on the set of formal matrix rings; see Item (a) in Sect. 4.1. It is clear that this action can be restricted to the set of the rings $M(n, R, s)$ of the considered form with fixed multiplier s. Every permutation τ also acts on each matrix $A = (a_{ij})$; $\tau A = (a_{\tau(i)\tau(j)})$. In particular, τ acts on multiplier matrices S. It follows from Lemma 4.3.2 that the orbits consist of matrices of the same canonical form. The number of orbits into which the set of multiplier matrices is decomposed under the action of permutations is equal to the number of representations of the integer n as sums of integers which are less than n.

If the ring $K = M(n, R, s)$ has the multiplier matrix S, then τS is a multiplier matrix of the ring τK for each permutation τ. Consequently, if two rings lie in the same orbit, then the corresponding multiplier matrices also lie in the same orbit. Of course, the converse is not true. The reason is that several series of multiplier matrices $S_1, \ldots, S_k$ can exist for this multiplier matrix S. Therefore, it is not clear how many ring orbits exist. In several special cases, the answer will be obtained in the next section.

For a given matrix ring $\begin{pmatrix} R & M \\ N & S \end{pmatrix}$ we introduced in Sect. 2.3 four construction methods for matrix rings of order > 2. We call these methods *constructions* K_1, K_2, K_3 and K_4. Now we assume that there is a ring $M(n, R, s) = L$ with multiplier matrices $S, S_1, \ldots, S_n$. We assume that the mentioned constructions are applied to L. We denote the obtained rings by L_1, L_2, L_3 and L_4, respectively. The multiplier matrices $S', S_1', \ldots, S_n', S_{n+1}'$ of each of these rings are obtained from matrices $S, S_1, \ldots, S_n$ with the use of a certain regularity. We consider the construction K_4 in more detail. We need to represent the ring L in block form as a matrix ring of order 2 (see Sect. 2.3), $L = \begin{pmatrix} P & M \\ N & S \end{pmatrix}$, where P is a matrix ring of order k, $1 \le k \le n - 1$; S, M and N are described in Sect. 2.3. For simplicity, we take $k = n - 1$. We apply to L the construction K_4 and obtain the matrix ring L_4 of order $n + 1$,

$$L_4 = \begin{pmatrix} L & \begin{pmatrix} M \\ S \end{pmatrix} \\ (N\ S) & S \end{pmatrix}.$$

By considering the structure of matrices in the ring L_4, we have that multiplier matrices $S', S_1', \ldots, S_n'$ of the ring L_4 are uniformly obtained from the corresponding matrices $S, S_1, \ldots, S_n$, and $S_{n+1}' = S_n'$. Namely, we have to add on the right of each matrix $S, S_1, \ldots, S_n$ the last column of this matrix; we have to add below each matrix $S, S_1, \ldots, S_n$ the last row of this matrix; in position $(n + 1, n = 1)$ we have to put the element appearing in position (n, n).

We formulate the following question, which will again appear at the end of Sect. 4.4. Let us act by a permutation τ of degree n on the ring L. Then we apply the construction K_4 to the ring τL and obtain the ring $(\tau L)_4$. What is the connection between the rings L_4 and $(\tau L)_4$? Is it possible to transfer one of these rings onto the second ring by some permutation?

Concluding this section, we consider the excluded values of the multiplier s: $s^2 = 1$ and $s^2 = s$. If $s^2 = 1$, then the element s is invertible and $M(n, R, s) \cong M(n, R)$ by Proposition 4.1.2. The case $s^2 = s$, where s is an idempotent, is more informative. Here is one specific example. Set $s_{iik} = 1 = s_{ikk}$ and $s_{ijk} = s$ for $i \ne j$ and $j \ne k$. Then $\Sigma = \{s_{ijk}\}$ is a multiplier system. Consequently, the ring $M(n, R, \Sigma)$ exists.

4.4 Classification of Formal Matrix Rings

We continue to study the considered in the previous section. We mainly focus on Problem (II) formulated in Sect. 4.1. We also finish the proof of Theorem 4.3.3 related to problem (I). All notations and agreements of Sect. 4.3 are preserved.

By Theorem 4.3.3, for any multiplier matrix T, there exists a ring $M(n, R, s)$ such that its multiplier matrix S coincides with T. Such a ring $M(n, R, s)$ is not necessarily unique, since it is possible that for the matrix S, there exist distinct families of multiplier matrices $S_1, \ldots, S_n$ corresponding to distinct multiplier systems Σ. In this situation, the classification problem (II) is to list all such rings $M(n, R, s)$ for fixed s.

Depending on the canonical form of the matrix S, we study the following two cases:

(1) the main diagonal of the matrix S consists of two blocks;

(2) all blocks on the main diagonal of the matrix S are of order 1.

In the first case, $S_1, \ldots, S_n$ are symmetric matrices and can be uniquely restored from the matrix S.

Similar to Sect. 4.3, we assume that the multiplier matrix of the ring $M(n, R, s)$ is the matrix S, unless stated otherwise.

Lemma 4.4.1 *Let us assume that we have the ring $M(n, R, s)$ and let $S, S_1, \ldots, S_n$ be multiplier matrices of the ring. The following conditions are equivalent.*

(1) *The canonical form of the matrix S contains exactly two blocks.*

(2) *For any three elements s_{iji}, s_{iki} and s_{jkj}, one of the possibilities (1) or (2) from Lemma 4.3.1 is realized.*

(3) *The matrices $S_1, \ldots, S_n$ are symmetric.*

Proof (1) $\Rightarrow$ (2). Let τ be an arbitrary permutation. If elements of the matrix S satisfy (2), then the same holds for elements of the matrix τS, and conversely. Therefore, we can assume that S is of the canonical form. Then it is easy to verify that (2) holds.

(2) $\Rightarrow$ (1). We again assume that S is of the canonical form. If we assume that the number of blocks is not less than 3, then we will soon obtain a contradiction.

(2) $\Rightarrow$ (3). We take any three elements s_{iji}, s_{iki} and s_{jkj}, where the subscripts i, j, k are pairwise distinct. By (2), either all these elements are equal to 1, or some two of them are equal to s, and the third element is equal to 1. In any case, it follows from relations (4.1.3) of Sect. 4.1 that $s_{ikj} = s_{jki}$.

(3) $\Rightarrow$ (2). If we assume that

$$s_{iji} = s_{iki} = s_{jkj} = s,$$

then we obtain that all six remaining multipliers in relation (4.1.4) of Sect. 4.1 are equal to each other. This contradicts relations (4.1.3) of Sect. 4.1.   □

Next we consider rings $M(n, R, s)$ with multiplier matrices which satisfy equivalent conditions (1)–(3) of Lemma 4.4.1. It is not difficult to prove the following assertion: each of the rows of the matrix S uniquely determines the remaining rows. We reformulate this assertion. For this purpose, we pass to the "additive" form in the matrices $S, S_1, \ldots, S_n$; we replace all elements by the corresponding exponents of the element s (as usual, we assume that $s^0 = 1$). The obtained matrices consist of 0s and 1s. (Such matrices are called (01) *matrices*.) We denote by S^+ the matrix constructed from the matrix S. We continue to denote elements of the matrix S^+ by s_{iji} or, more briefly, s_{ij}. For elements of the matrix S^+ in the field of two elements, assertion (2) of Lemma 4.4.1 has the form $s_{ij} + s_{ik} = s_{jk}$. This relation also holds if two of the subscripts coincide with each other.

The elements s_{ikj} with distinct i and j (elements of the matrix S_k) are uniquely determined by elements of the form s_{iji}. More precisely, elements of the main diagonal of the matrix S_k (this diagonal coincides with the kth row of the matrix S) determine all remaining elements of this matrix if we consider the relations (4.1.3) of Sect. 4.1. In the field of two elements, the relation $s_{ikj} = s_{ik} \cdot s_{jk}$ holds. It also remains true for coinciding subscripts. Thus, the matrix S completely determines the matrices S_k, $k = 1, \dots, n$.

We can offer a complete description of the rings $M(n, R, s)$ considered in Lemma 4.4.1.

We note that the ring $M(n, R)$ satisfies Item (3) of the following theorem. It corresponds to the case where all multipliers s_{ijk} are equal to $1 = s^0$. In Item (2), the matrix consisting of 1 s corresponds to $M(n, R)$.

Theorem 4.4.2 *There exists a one-to-one correspondence between the following three sets.*

(1) *The set of sequences of length $n - 1$ consisting of 0 s and 1 s.*
(2) *The set of multiplier matrices of order n which satisfy equivalent conditions (1) and (2) of Lemma 4.4.1.*
(3) *The set of rings $M(n, R, s)$ whose multiplier matrices satisfy conditions (1), (2), (3) of Lemma 4.4.1.*

Proof Substantially, the bijection between the sets (1) and (2) is already proved. Namely, if $T = (t_{ij})$ is some multiplier matrix in the set (2), then the sequence $t_{12}, \dots, t_{1n}$ corresponds to the matrix T (as mentioned above, we replace elements t_{ij} by the corresponding exponents of the element s; then we perform similar actions).

Conversely, assume that we have some sequence from the set (1). We insert a 0 at the beginning of this sequence. We take this extended sequence as the first row of the matrix (t_{ij}) of order n. The remaining elements of the matrix are obtained with the use of the relations $t_{1j} + t_{1k} = t_{jk}$ (in the field of two elements). We also have the relation $t_{ij} + t_{ik} = t_{jk}$ for all $i = 2, \dots, n$ and all j, k. Now we replace 0 (resp., 1) by 1 (resp., s) in this matrix. The constructed matrix is the matrix from the set (2). The obtained correspondence between sequences from (1) and matrices in (2) is a bijection.

Now we pass to a bijection between the sets (2) and (3); in fact, it already exists. If T is some matrix from the set (2), then by Theorem 4.3.3, there exists a ring $M(n, R, s)$ such that T is a multiplier matrix for $M(n, R, s)$. Now we observe that the coincidence of two rings of the form $M(n, R, \{s_{ijk}\})$ means that multiplication operations in these rings coincide. It follows from the definition of such rings that multiplier systems of these rings $M(n, R, \{s_{ijk}\})$ coincide. Therefore, it is clear that distinct rings $M(n, R, s)$ correspond to distinct matrices T.

Conversely, if a ring $M(n, R, s)$ is contained in the set (3), then the multiplier matrix S of this ring corresponds to the ring $M(n, R, s)$. It was noted earlier that multipliers s_{ijk} with distinct subscripts i, k are uniquely defined by multipliers of the form s_{iji}. In other words, multiplier matrices $S_1, \dots, S_n$ of the ring $M(n, R, s)$

are uniquely determined by the matrix S. Therefore, distinct rings $M(n, R, s)$ have distinct matrices S.   □

Corollary 4.4.3 *In Theorem 4.4.2, the number of rings in the set* (3) *is equal to* 2^{n-1}. *Consequently, this number does not depend on the ring R or the element s.*

Now we describe a general situation. Assume that we have an arbitrary set of central elements $\{s_{ijk} \mid i, j, k = 1, \ldots, n\}$ of the ring R which satisfies the relations (4.1.1) of Sect. 4.1. We have to verify whether it is a multiplier system. For this purpose, we particularly need to check the relation

$$s_{ijk} \cdot s_{ik\ell} = s_{ij\ell} \cdot s_{jk\ell} \tag{$*$}$$

for all pairwise distinct subscripts i, j, k, ℓ. It may be that for each such tetrad of subscripts, only their order is important. A similar situation occurs below. For convenience, we assume that $i = 1$, $j = 2$, $k = 3$ and $\ell = 4$. By considering all permutations of integers i, j, k, ℓ, we obtain that the number of the corresponding relations $(*)$ is equal to 24. It is convenient to arrange the relations as four series which have six relations in every series. We have the first series, in which subscripts i, j, k run through the set $\{1, 2, 3\}$:

$$s_{123} \cdot s_{134} = s_{124} \cdot s_{234}, \; s_{321} \cdot s_{314} = s_{324} \cdot s_{214},$$
$$s_{312} \cdot s_{324} = s_{314} \cdot s_{124}, \; s_{213} \cdot s_{234} = s_{214} \cdot s_{134}, \tag{$**$}$$
$$s_{231} \cdot s_{214} = s_{234} \cdot s_{314}, \; s_{132} \cdot s_{124} = s_{134} \cdot s_{324}.$$

There are three additional series for values of subscripts $\{1, 2, 4\}$, $\{1, 3, 4\}$ and $\{2, 3, 4\}$.

Now we pass to the case (2) mentioned at the beginning of this section. Namely, we study another situation occurring with the ring $M(n, R, s)$, where all multipliers s_{iji}, $i \neq j$, are equal to s. In other words, the multiplier matrix of such a ring has blocks of order 1 on the main diagonal; this matrix is already in the canonical form. In particular, we finish the proof of Theorem 4.3.3. The ring $M(2, R, s)$ is a ring of this type; therefore, we assume that $n \geq 3$.

Theorem 4.4.4 *Let $T = (t_{ij})$ be a matrix of order $n \geq 3$ such that 1 stands on the main diagonal and the element s appears in the remaining positions. There exists a ring $M(n, R, s)$ such that T is a multiplier matrix of $M(n, R, s)$. Any two such rings are mapped onto each other by permutations, that is they are contained in the same orbit.*

Proof From the remarks after Corollary 4.4.3, it is clear what values should be given to the multipliers s_{ijk} to obtain the required ring $M(n, R, s)$. As always, $s_{iik} = 1 = s_{ikk}$. Then we assume that $s_{iji} = t_{ij}$. For any pairwise distinct subscripts i, j, k, we assume that $s_{ijk} = 1$ if the permutation (i, j, k) is even and $s_{ijk} = s$ for an odd permutation (i, j, k). We verify that the set $\Sigma = \{s_{ijk} \mid i, j, k = 1, \ldots, n\}$ is a multiplier system and identities (4.1.1) of Sect. 4.1 hold.

If two subscripts coincide in some identity (4.1.1) (this is the identity $(*)$ above), then it is turned into an identity of the form (4.1.3) or (4.1.4) of Sect. 4.1. However, these identities hold by the choice of multipliers s_{ijk}. In particular, the property holds in the case $n = 3$. Next we assume that $n \geq 4$. It remains to verify the relation $(*)$ for any pairwise distinct subscripts i, j, k, ℓ. Here we are in the situation described after Theorem 4.3.3, since only the arrangement of the integers i, j, k, ℓ is important in the check. Therefore, it is sufficient to verify that all 24 relations of the form $(**)$ hold, which is directly verified.

Thus, Σ is a multiplier system; consequently, there exists a ring $M(n, R, \Sigma)$ with multiplier matrix T.

Remark We interrupt the proof of Theorem 4.4.4; later, we return to the remaining assertion about the orbit.

Conversely, we can assume that $s_{ijk} = s$ for even permutations (i, j, k) and $s_{ijk} = 1$ for odd permutations (i, j, k). Then we also obtain a ring with multiplier matrix T. We denote by L_0 and L_1 the corresponding rings. The permutation $\sigma = \begin{pmatrix} 1 & 2 & \dots n \\ n & n-1 & \dots 1 \end{pmatrix}$ maps L_0 onto L_1 (we use the fact that the permutations (i, j, k) and $(\sigma(i), \sigma(j), \sigma(k))$ have opposite parities).

Thus, we have obtained two rings L_0 and L_1 with multiplier matrix T. Are there more such rings? For $n = 3$, the answer is negative; this follows from the following paragraph.

We take $n = 4$. It directly follows from identities (4.1.3) and (4.1.4) of Sect. 4.1 that the multipliers $s_{123}, s_{312}, s_{231}$ from the relations $(**)$ coincide, and the same holds for $s_{321}, s_{213}, s_{132}$. In addition, elements of these triples have to take opposite values (from 1 and s). We have a similar situation with the first elements of the remaining 18 relations. The first elements of the relations $(**)$ coincide with all elements presented in these relations. Therefore, we obtain that the given ring $M(4, R, s)$ is uniquely determined by the vector (c_1, c_2, c_3, c_4) of length 4, consisting of 0s and 1s. Here $c_1 = 0$ if $s_{123} = 1$, $c_1 = 1$ if $s_{123} = s$, and so on. The rings L_0 and L_1 meet vectors $(0, 0, 0, 0)$ and $(1, 1, 1, 1)$, respectively. By applying cycles $(1\,2\,4\,3), (1\,3\,2\,4)$ and $(1\,4\,2\,3)$ to the rings L_0 and L_1, we obtain four additional rings $M(4, R, s)$, to which the vectors $(0, 1, 1, 0), (1, 0, 0, 1), (0, 0, 1, 1), (1, 1, 0, 0)$ correspond. The permutation $(1\,4)(2\,3)$ maps L_0 onto L_1, and conversely. There do not exist rings $M(4, R, s)$ corresponding to the vectors $(0, 1, 0, 1), (0, 1, 1, 1)$, $(1, 0, 1, 1), (1, 1, 0, 1), (0, 0, 0, 1)$; to verify this, it is sufficient to consider the relations $(**)$. Therefore, there exist six rings $M(4, R, s)$ with this multiplier matrix T. These rings are transformed onto each other by permutations; they are contained in the same orbit. In addition, we note that

$$L_1 = L_0^t, \quad L_{(0,0,1,1)} = L_{(1,1,0,0)}^t,$$
$$L_{(0,1,1,0)} = L_{(1,0,0,1)}^t;$$

see Item (d) of Sect. 4.1.

Now we assume that $n \geq 5$. We already have two rings L_0 and L_1 such that T is a multiplier matrix for L_0 and L_1. Now let $M(n, R, s)$ be one more such a ring. We show that it is mapped by some permutation onto L_0 or L_1.

We fix arbitrary pairwise distinct subscripts i, j, k, ℓ. The elements of the matrices in $M(n, R, s)$ appearing in positions

$$(i,i),\ (i,j),\ (i,k),\ (i,\ell),\ (j,i),\ (j,j),\ (j,k),\ (j,\ell),$$
$$(k,i),\ (k,j),\ (k,k),\ (k,\ell),\ (\ell,i),\ (\ell,j),\ (\ell,k),\ (\ell,\ell),$$

form a ring $M(4, R, s)$. We have just proved that there exists a permutation σ of degree 4 which maps the ring $M(4, R, s)$ onto the ring L_0. By fixing those integers which are not equal to i, j, k, ℓ, we regard σ as a permutation of degree n. We apply σ to the ring $M(n, R, s)$. By repeating this action several times, we obtain a permutation of degree n which maps the ring $M(n, R, s)$ onto L_0. Thus, the proof of Theorem 4.4.4 is complete.   □

We use this to write the above-considered multiplier matrices S of rings $M(n, R, s)$ for $n = 2, 3, 4$. We omit the matrix which consists of 1 s and corresponds to the ring $M(n, R)$.

$n = 2$. $S = \begin{pmatrix} 1 & s \\ s & 1 \end{pmatrix}$.

$n = 3$. Two orbits of multiplier matrices and two corresponding orbits of rings.

The first orbit: $S = \begin{pmatrix} 1 & s & s \\ s & 1 & s \\ s & s & 1 \end{pmatrix}$.

The orbit of rings consists of the rings L_0 and L_1.

The second orbit:

$$\begin{pmatrix} 1 & s & s \\ s & 1 & 1 \\ s & 1 & 1 \end{pmatrix}, \quad \begin{pmatrix} 1 & 1 & s \\ 1 & 1 & s \\ s & s & 1 \end{pmatrix}, \quad \begin{pmatrix} 1 & s & 1 \\ s & 1 & s \\ 1 & s & 1 \end{pmatrix}.$$

The orbit of rings contains three rings.

$n = 4$. There are three orbits of multiplier matrices and three corresponding orbits of rings. The first orbit: $S = \begin{pmatrix} 1 & s & s & s \\ s & 1 & s & s \\ s & s & 1 & s \\ s & s & s & 1 \end{pmatrix}$.

The orbit of rings consists of six rings.

The second orbit:

$$\begin{pmatrix} 1 & s & s & s \\ s & 1 & 1 & 1 \\ s & 1 & 1 & 1 \\ s & 1 & 1 & 1 \end{pmatrix}, \quad \begin{pmatrix} 1 & 1 & 1 & s \\ 1 & 1 & 1 & s \\ 1 & 1 & 1 & s \\ s & s & s & 1 \end{pmatrix}, \quad \begin{pmatrix} 1 & s & 1 & 1 \\ s & 1 & s & s \\ 1 & s & 1 & 1 \\ 1 & s & 1 & 1 \end{pmatrix}, \quad \begin{pmatrix} 1 & 1 & s & 1 \\ 1 & 1 & s & 1 \\ s & s & 1 & s \\ 1 & 1 & s & 1 \end{pmatrix}.$$

The third orbit:

$$\begin{pmatrix} 1 & 1 & s & s \\ 1 & 1 & s & s \\ s & s & 1 & 1 \\ s & s & 1 & 1 \end{pmatrix}, \quad \begin{pmatrix} 1 & s & s & 1 \\ s & 1 & 1 & s \\ s & 1 & 1 & s \\ 1 & s & s & 1 \end{pmatrix}, \quad \begin{pmatrix} 1 & s & 1 & s \\ s & 1 & s & 1 \\ 1 & s & 1 & s \\ s & 1 & s & 1 \end{pmatrix}.$$

The second orbit of rings contains 4 rings, the third orbit of rings contains 3 rings.

The rings L_0 are examples of rings of crossed matrices [11], which are quite useful for the structure theory of some Artinian rings.

We return to Lemma 4.3.1. This lemma concerns possible values of multipliers in the triple $s_{iji}, s_{iki}, s_{jkj}$. We considered the situation where only possibilities (1) and (2) are realized, and then when only (3) is realized. A certain structure of the canonical form of multiplier matrices corresponds to these situations. It remains to consider the case where all three possibilities (1), (2), (3) can appear. This is equivalent to the property that the canonical form of this multiplier matrix has at least three blocks on the main diagonal, and there are blocks of order exceeding 1.

How can we obtain all rings $M(n, R, s)$ such that the canonical form of the multiplier matrix has a similar structure? For $n = 2$ and $n = 3$, such rings do not exist. The proof of Theorem 4.3.3 provides a practical method of constructing the required rings for $n \geq 4$. Namely, we have to take rings $M(n, R, s)$, beginning with $n = 3$, and apply them to the construction K_4 from Sect. 2.3. Before this, we have to apply some permutation to pass to the ring $M(n, R, s)$ whose multiplier matrix is of the canonical form. From the rings $M(n, R, s)$, we have to exclude those rings whose multiplier matrices have the canonical form containing exactly two blocks on the main diagonal. As a result, we obtain rings $M(n+1, R, s)$ with the required form of the multiplier matrix. Then we apply all possible permutations to these rings. After this, we will have the required rings $M(n + 1, R, s)$. If the blocks of the multiplier matrix of the ring $M(n, R, s)$ have orders $k_1, \ldots, k_t$ with $t \geq 3$, then blocks of the corresponding ring $M(n + 1, R, s)$ have orders $k_1, \ldots, k_{t-1}, k_t + 1$.

We illustrate the above construction method for $n = 4$. We take the ring $M(3, R, s)$ denoted above by L_0. This ring has the multiplier matrix $\begin{pmatrix} 1 & s & s \\ s & 1 & s \\ s & s & 1 \end{pmatrix}$. By applying the construction K_4 to the ring L_0, we obtain the ring $M(4, R, s)$ with multiplier matrix

$$\begin{pmatrix} 1 & s & s & s \\ s & 1 & s & s \\ s & s & 1 & 1 \\ s & s & 1 & 1 \end{pmatrix},$$

for which $s_{121} = s_{131} = s_{232} = s_{141} = s$, $s_{343} = 1$. By considering the position of the 1 above the main diagonal, we denote the obtained ring by L_{34}. We apply permutations $(3\,1)(4\,2)$, $(1\,3\,4)(2)$, $(1\,3)(2)(4)$, $(2\,3\,4)(1)$, $(2\,3)(1)(4)$ to the ring L_{34} and obtain rings $L_{12}, L_{13}, L_{14}, L_{23}, L_{24}$, respectively. Above the main diagonal, multiplier matrices of these rings have a 1 in positions $(1, 2)$, $(1, 3)$, $(1, 4)$, $(2, 3)$ and $(2, 4)$, respectively.

We can do the same, beginning with the ring $L_1 = M(3, R, s)$. As a result, we obtain rings $L'_{12}, L'_{13}, L'_{14}, L'_{23}, L'_{24}$ and L'_{34}. We recall that the ring L_0 is mapped onto the ring L_1 by some permutation (for example, $(1\,4)(2\,3)$). The ring L_{34} is mapped onto the ring L'_{34} by the permutation $(1\,2)(3\,4)$; this provides a partial answer to the question formulated at the end of Sect. 4.3. Thus, all 12 rings

$$L_{12}, \ L_{13}, \ L_{14}, \ L_{23}, \ L_{24}, \ L_{34},$$
$$L'_{12}, \ L'_{13}, \ L'_{14}, \ L'_{23}, \ L'_{24}, \ L'_{34}$$

are contained in the same orbit. In addition, we note that $L_1 = L'_0$ and $L'_{ij} = L^t_{ij}$; see Item (d) of Sect. 4.1.

For $n = 4$, one more orbit can be added to the above list of multiplier matrices,

$$
\begin{pmatrix} 1 & 1 & s & s \\ 1 & 1 & s & s \\ s & s & 1 & s \\ s & s & s & 1 \end{pmatrix}, \quad
\begin{pmatrix} 1 & s & 1 & s \\ s & 1 & s & s \\ 1 & s & 1 & s \\ s & s & s & 1 \end{pmatrix}, \quad
\begin{pmatrix} 1 & s & s & 1 \\ s & 1 & s & s \\ s & s & 1 & s \\ 1 & s & s & 1 \end{pmatrix},
$$

$$
\begin{pmatrix} 1 & s & s & s \\ s & 1 & 1 & s \\ s & 1 & 1 & s \\ s & s & s & 1 \end{pmatrix}, \quad
\begin{pmatrix} 1 & s & s & s \\ s & 1 & s & 1 \\ s & s & 1 & s \\ s & 1 & s & 1 \end{pmatrix}, \quad
\begin{pmatrix} 1 & s & s & s \\ s & 1 & s & s \\ s & s & 1 & 1 \\ s & s & 1 & 1 \end{pmatrix}.
$$

The corresponding orbit of rings contains 12 rings of the form L_{ij} and L'_{ij}. It remains unclear whether there are other rings in this orbit.

4.5 The Isomorphism Problem

The following result can be directly obtained from Lemma 2.4.5.

Lemma 4.5.1 *For any formal matrix ring K over the ring R, the relation $C(K) = \{rE \mid r \in C(R)\}$ holds.*

In Sect. 4.1, the isomorphism problem (III) is formulated. We now consider it for formal matrix rings $M(n, R, \{s_{ijk}\})$ such that for $i \neq j$ and $j \neq k$, every multiplier s_{ijk} is equal to s^m for some $m \geq 1$, where s is a fixed central element of the ring R. Similar rings were mentioned at the beginning of Sect. 4.3. Here we denote such rings by $M(n, R, s)$; not to be confused with the symbol $M(n, R, s)$ from Sects. 4.3 and 4.4, where $M(n, R, s)$ denotes a ring such that $s_{ijk} = 1$ or $s_{ijk} = s$.

In what follows, $M(n, R, 0)$ denotes a ring $M(n, R, \{s_{ijk}\})$ such that all multipliers s_{ijk} are equal to zero, except for s_{iik} and s_{ikk}. Further, T is some ring and $M(n, T, \{t_{ijk}\})$ is an arbitrary formal matrix ring over T.

We recall the following definition. A ring is said to be *normal* if all its idempotents are central.

Lemma 4.5.2 *Let R be a normal ring. If $M(n, R, 0) \cong M(n, T, \{t_{ijk}\})$, then all multipliers t_{ijk} are equal to zero, except for the cases where $i = j$ or $j = k$.*

Proof Set $K_1 = M(n, R, 0)$ and $K_2 = M(n, T, \{t_{ijk}\})$. We fix some ring isomorphism $f: K_1 \to K_2$. Let I be the ideal (I_{ij}) of the ring K_1, where $I_{ii} = 0$ and $I_{ij} = R$ for $i \neq j$; see Proposition 4.2.4. Then $I^2 = 0$, whence $(f(I))^2 = 0$. We assume that there exists a non-zero multiplier t_{ikj} such that $i \neq k$ and $k \neq j$. Then $E_{ik}E_{kj} = t_{ikj}E_{ij} \neq 0$, where E_{ij} is a matrix unit; see Sect. 4.1. Consequently, the matrices E_{ik} and E_{kj} cannot simultaneously be contained in $f(I)$. For definiteness, let $E_{ik} \notin f(I)$. Then

$$E_{ii}(E_{ii} + E_{ik}) - (E_{ii} + E_{ik})E_{ii} = E_{ik} \notin f(I).$$

This shows that the idempotent $(E_{ii} + E_{ik}) + f(I)$ of the factor ring $K_2/f(I)$ is not central in $K_2/f(I)$. Therefore, the ring $K_2/f(I)$ is not normal. On the other hand, the ring $K_2/f(I)$ is normal, since

$$K_2/f(I) \cong K_1/I \cong R \oplus \ldots \oplus R$$

is the finite direct product of normal rings. This is a contradiction.   □

Now let s and t be two non-zero central elements of the ring R (by Lemma 4.5.2, we can assume that $s \neq 0$ and $t \neq 0$). In addition, let $s^k \neq s^\ell$ for any distinct non-negative k and ℓ. Further, let $M(n, R, \{s_{ijk}\})$, $M(n, R, \{t_{ijk}\})$ be the rings considered at the beginning of this section; namely, every multiplier s_{ijk} is a positive integral degree of the element s and every multiplier t_{ijk} is a positive integral degree of the element t. We also assume that at least one of the multipliers s_{ijk} is equal to s and at least one of the multipliers t_{ijk} is equal to t. In addition, we assume that the multiplier systems s_{ijk} and t_{ijk} are "similar" in the following sense:

$$s_{ijk} = s^m \quad \Longleftrightarrow \quad t_{ijk} = t^m$$

for all multipliers s_{ijk} and t_{ijk}. In accordance with our agreement, we denote the considered rings by $M(n, R, s)$ and $M(n, R, t)$, respectively.

We recall that $J(R)$, $U(R)$ and $Z(R)$ denote the Jacobson radical, the group of invertible elements and the set of all (left or right) zero-divisors of the ring R, respectively.

Theorem 4.5.3 *Let R be a commutative ring with $Z(R) \subseteq J(R)$. The rings $M(n, R, s)$ and $M(n, R, t)$ are isomorphic if and only if $t = v\alpha(s)$, where v is an invertible element in R and α is an automorphism of the ring R.*

Proof Set $K_1 = M(n, R, s)$ and $K_2 = M(n, R, t)$. Assume that we have a ring isomorphism $f: K_1 \to K_2$. The isomorphism f induces an isomorphism $C(K_1) \to C(K_2)$ between the centers of these rings. We consider the action of this isomorphism in detail. We take an arbitrary element $a \in R$. By Lemma 4.5.1, $aE \in C(K_1)$.

Consequently, $f(aE) \in C(K_2)$. Then $f(aE) = bE$ for some $b \in R$. We obtain that f induces the automorphism α of the ring R with $\alpha(a) = b$. Thus, we have $f(aE) = \alpha(a)E, a \in R$.

Now we take the ideal $(sE)K_1 = \begin{pmatrix} sR \ldots sR \\ \ldots\ldots\ldots \\ sR \ldots sR \end{pmatrix}$ of the ring K_1. The image of this ideal under the action f is the ideal $f((sE)K_1)$. We have the relations

$$f((sE)K_1) = f(sE)K_2 = (\alpha(s)E)K_2 = \begin{pmatrix} \alpha(s)R \ldots \alpha(s)R \\ \ldots \quad \ldots \quad \ldots \\ \alpha(s)R \ldots \alpha(s)R \end{pmatrix}.$$

The isomorphism f induces an isomorphism of the factor rings $K_1/(sE)K_1 \rightarrow K_2/(\alpha(s)E)K_2$. The first factor ring is the matrix ring $M(n, R/sR, 0)$; see the paragraph before Lemma 4.5.2. The second factor ring is the matrix ring $M(n, R/\alpha(s)R, \bar{t})$, where $\bar{t} = t + \alpha(s)R$. Thus, the multipliers of this ring are residue classes $t_{ijk} + \alpha(s)R$. By applying Lemma 4.5.2, we obtain that $\bar{t} = 0$ or $t \in \alpha(s)R$ (where t is one of the multipliers t_{ijk}). Consequently, $t = \alpha(s)x$ for some element $x \in R$. By considering the inverse isomorphism to f, we obtain that $s = \alpha^{-1}(t)y, y \in R$. Then we have

$$t = \alpha(s)x = t\alpha(y)x, \qquad t(1 - \alpha(y)x) = 0.$$

By assumption, we obtain that

$$1 - \alpha(y)x \in J(R), \quad 1 - (1 - \alpha(y)x) = \alpha(y)x \in U(R).$$

Consequently, x is an invertible element. Therefore, $t = v\alpha(s)$, where v is an invertible element and α is an automorphism of the ring R.

Now we assume that $t = v\alpha(s)$, where v is an invertible element and α is an automorphism of the ring R. We show that the rings K_1 and K_2 are isomorphic. Firstly, it follows from Item (b) of Sect. 4.1 that there is an isomorphism $M(n, R, s) \cong M(n, R, \alpha(s))$. The multiplier system of the second ring is $\{\alpha(s_{ijk}) \mid i, j, k = 1, \ldots, n\}$. We can assume that α is the identity automorphism. We take the set $\{v_{ijk} \mid i, j, k = 1, \ldots, n\}$, where $v_{ijk} = v^m$ provided $s_{ijk} = s^m$. This set is a multiplier system. Therefore, we have the ring $M(n, R, vs)$ with multiplier system $\{v_{ijk}s_{ijk} \mid i, j, k = 1, \ldots, n\}$. By Item (c) of Sect. 4.1, there exists an isomorphism $M(n, R, vs) \cong M(n, R, s)$. It follows from the relations $t = vs$ and our agreement about "similarity" of multiplier systems that $t_{ijk} = v_{ijk}s_{ijk}$ for all i, j, k. Therefore, $M(n, R, vs) = M(n, R, t)$ and $M(n, R, s) \cong M(n, R, t)$. $\square$

We remark that $Z(R) \subseteq J(R)$ if R is either a domain or a local ring.

Corollary 4.5.4 *Let R be a commutative ring which is either a domain or a local ring. The rings $M(n, R, s)$ and $M(n, R, t)$ are isomorphic if and only if $t = v\alpha(s)$, where v is an invertible element and α is an automorphism of the ring R.*

In the papers of Krylov [70] and Tang, Li, Zhou [104], some isomorphism theorems are proved for formal matrix rings of order 2 (in particular, Corollary 4.5.5). Tang and Zhou [106] proved Theorem 4.5.3 for the rings mentioned in Sect. 4.3. In this case, the restrictions on elements s and t, listed before Theorem 4.5.3, are not required. These restrictions are also unnecessary for $n = 2$ (the multiplication rules of matrices in such a ring are presented after Proposition 4.1.2).

Corollary 4.5.5 ([70]) *Let R be a commutative ring and s, t two elements of this ring such that at least one of them is not a zero divisor. The rings $M(2, R, s)$ and $M(2, R, t)$ are isomorphic if and only if $t = v\alpha(s)$, where v is an invertible element and α is an automorphism of the ring R.*

Abyzov and Tapkin [3] study rings $M(3, R, \Sigma)$ such that S_1, S_2 and S_3 are symmetric matrices; $s_{ikj} = s_{jki}$ for all $i, j, k = 1, 2, 3$ (see the end of Sect. 4.1 for more about matrix rings of order 3). In particular, the authors obtained several results about the isomorphism problem for such rings; see also [4]. In addition, Abyzov and Tapkin introduced multiplier systems of a more general form (as compared with the case of $n = 3$) as follows. Let $s_1, \ldots, s_n$ be arbitrary central elements of the ring R. For all $i, j, k = 1, \ldots, n$, we set

$$s_{ijk} = \begin{cases} 1, & \text{if } i = j \text{ or } j = k; \\ s_j, & \text{if } i, j, k \text{ are pairwise distinct}; \\ s_i s_j, & \text{if } i = k, \text{ but } i \neq j. \end{cases}$$

One can directly verify that $\Sigma = \{s_{ijk}\}$ is a multiplier system, since Σ satisfies relations (4.1.1) of Sect. 4.1. Consequently, the ring $M(n, R, \Sigma)$ exists. Further, Abyzov and Tapkin [3] extend results about the isomorphism problem, obtained for the rings $M(3, R, \Sigma)$, to the above-mentioned rings $M(n, R, \Sigma)$. The class of such rings $M(n, R, \Sigma)$ contains the rings of Tang and Zhou from the beginning of Sect. 4.3.

If s and t are arbitrary invertible elements, then any two rings $M(n, R, s)$ and $M(n, R, t)$, mentioned at the beginning of this section, are isomorphic by Proposition 4.1.2. Therefore, we can assume that one of these elements (for example, s) is not invertible. In this case, if R is a domain, then $s^k \neq s^\ell$ for $k \neq \ell$. This is realized in the following two examples. Therefore, we assume in these examples that the corresponding multiplier systems $\{s_{ijk}\}$ and $\{t_{ijk}\}$ satisfy only two of the conditions formulated before Theorem 4.5.3.

Example 4.5.6 Let $s, t \in \mathbb{Z}$. The isomorphism $M(n, \mathbb{Z}, s) \cong M(n, \mathbb{Z}, t)$ exists if and only if $t = \pm s$.

Proof First of all, we have to consider Lemma 4.5.2 and the remark before Example 4.5.6. Then the assertion follows from Theorem 4.5.3, since the ring $\mathbb{Z}$ has only the identity automorphism and $U(\mathbb{Z}) = \{1, -1\}$. $\square$

Example 4.5.7 Let R be a commutative domain and i, j two positive integers. If $M(n, R[x], x^i) \cong M(n, R[x], x^j)$ or $M(n, R[[x]], x^i) \cong M(n, R[[x]], x^j)$, then $i = j$.

Proof Let V be one of the rings $R[x]$, $R[[x]]$. We assume that $i < j$. By Theorem 4.5.3, we have $x^i = v(x)\alpha(x^j) = v(x)\alpha(x)^j$, where $v(x)$ is an invertible element of the ring V and α is an automorphism of the ring V. Then we have $\alpha(x) = x^k(a_0 + a_1 x + \ldots)$, where $k \geq 0$ and $a_0 \neq 0$. Then $x^i = x^{jk} v(x)(a_0 + a_1 x + \ldots)^j$. However, $i < j$. Therefore, $k = 0$ and we have the relation $x^i = v(x)(a_0 + a_1 x + \ldots)^j$. Since $a_0 \neq 0$, we have that $v(x)$ is divisible by x, which is impossible. $\qquad\square$

4.6 Determinants of Formal Matrices

In Sects. 4.6 and 4.7, R is a commutative ring, and K denotes some formal matrix ring of order n over the ring R with a multiplier system $\{s_{ijk} \mid i, j, k = 1, \ldots, n\}$, $K = M(n, R, \{s_{ijk}\})$.

We recall the multiplication rule of matrices in the ring K. Let $A = (a_{ij})$, $B = (b_{ij})$ and $AB = (c_{ij})$. Then $c_{ij} = \sum_{k=1}^n s_{ikj} a_{ik} b_{kj}$. We use several times the relations from Sect. 4.1 concerning the interrelations between the multipliers s_{ijk}. For convenience, we partially repeat these relations. First of all, these are the main identities (4.1.1):

$$s_{iik} = 1 = s_{ikk}, \quad s_{ijk} \cdot s_{ik\ell} = s_{ij\ell} \cdot s_{jk\ell}.$$

From these identities, some other identities follow:

$$s_{iji} = s_{jij}, \quad s_{iji} = s_{ij\ell} \cdot s_{ji\ell} = s_{\ell ij} \cdot s_{\ell ji}.$$

In addition, for any $i, j = 1, \ldots, n$, we defined the element t_{ij} which is equal to $s_{ij\ell}$ for some $\ell = 1, \ldots, n$. We have the relations

$$t_{ij} \cdot t_{ji} = s_{iji}, \quad t_{ij} \cdot t_{jk} = t_{ik} \cdot s_{ijk}.$$

Now we define the notion of the determinant of an arbitrary matrix in K and prove that this determinant has properties which are similar to the main properties of the ordinary determinant of matrices over the ring $M(n, R)$. We use properties of this ordinary determinant without additional comments.

In several cases, determinants of matrices with entries in the ring K appear in the papers [3, 43, 44, 105, 106]. The paper [69] contains a general approach to the notion of the determinant of an arbitrary formal matrix of order 2.

We introduce some transformations of rows and columns of matrices in K.

(a). We can multiply rows of the matrix A by elements of R. Therefore, we can speak of common factors of elements of rows.

We have a homomorphism

$$\varphi_{ijk} \colon R_{ij} \otimes_{R_j} R_{jk} \to R_{ik}, \quad \varphi_{ijk}(x \otimes y) = s_{ijk} xy = x \circ y;$$

see Sects. 2.3 and 4.1. The symbol ∘ can be considered as an operation in R; however, its result depends on the subscripts i, j and k. To emphasize that the element r of the ring R is used as an element of R_{ij}, we add to this element subscripts i and j. Namely, we assume that $r_{ij} = r$. Let $A_1, \ldots, A_n$ be rows of the matrix A. Then $r_{ij} \circ A_j$ denotes the row vector $(r_{ij} \circ a_{j1}, \ldots, r_{ij} \circ a_{jn})$. In fact, we deal with some R-module $(R, \ldots, R)$.

(b). We can multiply (in the sense of the operation ∘) the jth row of the matrix A by the element r_{ij} and add it to the ith row. We briefly write such a transformation as $r_{ij} \circ A_j + A_i$.

(c). We can replace the ith row of the matrix A by the row $1_{ij} \circ A_j = (s_{ij1}a_{j1}, \ldots, s_{ijn}a_{jn})$; we also replace the jth row of the matrix A by the row $1_{ji} \circ A_i = (s_{ji1}a_{i1}, \ldots, s_{jin}a_{in})$. The transformation from the matrix A to the obtained matrix is called a *permutation* of the ith row and jth row of the matrix A. We can perform similar transformations over columns of the matrix A.

We say that the ith row and the jth row of the matrix A are *proportional* if $A_j = r_{ji} \circ A_i$ or $A_i = r_{ij} \circ A_j$ for some elements $r_{ji}, r_{ij} \in R$.

Let η be one of the homomorphisms

$$M(n, R, \{s_{ijk}\}) \to M(n, R)$$

defined before Proposition 4.1.2. It acts with the use of the relation $(a_{ij}) \to (t_{ij}a_{ij})$ where $t_{ij} = s_{ij\ell}$ for some fixed integer $\ell = 1, \ldots, n$.

We denote the ordinary determinant of the matrix $C \in M(n, R)$ by $|C|$. For each matrix A from the ring K, we set $d(A) = |\eta A|$. We call the element $d(A)$ the *determinant* of the matrix A in the ring K; the mapping $d \colon K \to R$, $A \to d(A)$, is called the *determinant of the ring K*.

We present an equivalent method of defining determinants of matrices in K. First, we return to the operation ∘. If $x_{ij}, x_{jk} \in R$, then $x_{ij} \circ x_{jk} = s_{ijk}x_{ij}x_{jk}$ by our agreement. We add subscripts i, k to the element $x_{ij}x_{jk}$.

Now assume that we have elements $a_{i_1 i_2}, \ldots, a_{i_{k-1} i_k}$ of the ring R. The expression

$$a_{i_1 i_2} \circ a_{i_2 i_3} \circ \ldots \circ a_{i_{k-1} i_k} \qquad (4.6.1)$$

has a precise meaning. Indeed, any placement of parentheses in (4.6.1) is valid; ∘ is an associative operation. This can be proved by induction on the number of elements. For $k = 3$, the assertion follows from the main identities mentioned at the beginning of this section.

We can assign a value to the expression (4.6.1) with the use of the tensor product $R_{i_1 i_2} \otimes_R \ldots \otimes_R R_{i_{k-1} i_k}$. The homomorphisms φ_{ijk} induce the homomorphism φ from this product into R. Then

$$a_{i_1 i_2} \circ \ldots \circ a_{i_{k-1} i_k} = \varphi(a_{i_1 i_2} \otimes \ldots \otimes a_{i_{k-1} i_k}).$$

We can also proceed as follows. As earlier, let E_{ij} be a matrix unit. Then $E_{ij} E_{jk} = s_{ijk} E_{ik}$. If

$$(a_{i_1 i_2} E_{i_1 i_2}) \cdot \ldots \cdot (a_{i_{k-1} i_k} E_{i_{k-1} i_k}) = c E_{i_1 i_k}, \quad c \in R,$$

then $a_{i_1 i_2} \circ \ldots \circ a_{i_{k-1} i_k} = c$.

Assume that we have an element $a_{i_k i_1}$. The subscripts of elements $a_{i_1 i_2}, \ldots, a_{i_{k-1} i_k}$, $a_{i_k i_1}$ form a cycle; we denote it by σ. We write the cycle, beginning with another element. In this case, the expression $a_{i_1 i_2} \circ \ldots \circ a_{i_{k-1} i_k} \circ a_{i_k i_1}$ is equal to the corresponding expression for another form of the cycle σ (we have to consider the relation $s_{iji} = s_{jij}$).

Now we can give a precise value to the expression $a_{1 i_1} \circ \ldots \circ a_{n i_n}$ under the condition that the second subscripts of multipliers form a permutation of the integers $1, \ldots, n$. For this purpose, we write the permutation

$$\tau = \begin{pmatrix} 1 & 2 & \ldots & n \\ i_1 & i_2 & \ldots & i_n \end{pmatrix}$$

as a product of independent cycles $\sigma_1, \ldots, \sigma_m$. If $c_1, \ldots, c_m$ are products (in the sense of the operation $\circ$) of elements, whose subscripts belong to cycles $\sigma_1, \ldots, \sigma_m$, respectively, then we assume that

$$a_{1 i_1} \circ \ldots \circ a_{n i_n} = c_1 \cdot \ldots \cdot c_m.$$

We assert that

$$d(A) = \sum_{n!} (-1)^q a_{1 i_1} \circ \ldots \circ a_{n i_n},$$

where q is the number of inversions in the permutation $i_1, \ldots, i_n$. Since

$$d(A) = |\eta A| = \sum_{n!} (-1)^q t_{1 i_1} a_{1 i_1} \cdot \ldots \cdot t_{n i_n} a_{n i_n},$$

it is sufficient to verify that the corresponding summands of two sums are equal to each other. We take two such summands and decompose the permutation of subscripts of multipliers of these summands into a product of independent cycles. We show that products of elements whose subscripts form some cycle are equal to each other. We take one of the cycles $(i_1 i_2 \ldots i_k)$. The product $a_{i_1 i_2} \circ \ldots \circ a_{i_k i_1}$ is equal to

$$s_{i_1 i_2 i_1} \cdot s_{i_2 i_3 i_1} \cdot \ldots \cdot s_{i_{k-1} i_k i_1} \cdot a_{i_1 i_2} \cdot \ldots \cdot a_{i_k i_1}.$$

The corresponding product for the determinant ηA is

$$t_{i_1 i_2} \cdot \ldots \cdot t_{i_k i_1} \cdot a_{i_1 i_2} \cdot \ldots \cdot a_{i_k i_1}.$$

By induction on the length of the cycle, we prove that

$$t_{i_1 i_2} \cdot \ldots \cdot t_{i_k i_1} = s_{i_1 i_2 i_1} \cdot \ldots \cdot s_{i_{k-1} i_k i_1}.$$

If $k = 2$, then $t_{i_1 i_2} t_{i_2 i_1} = s_{i_1 i_2 i_1}$. Let $k \geq 3$. We have $t_{i_{k-1} i_k} t_{i_k i_1} = t_{i_{k-1} i_1} s_{i_{k-1} i_k i_1}$; then we apply the induction hypothesis. As a result, we obtain the relation

$$d(A) = \sum_{n!} (-1)^q a_{1 i_1} \circ \ldots \circ a_{n i_n}.$$

We can say that the *complete development formula* holds for the determinant $d(A)$. As a corollary, we obtain that the first definition of the determinant $d(A)$ does not depend on the choice of the homomorphism η.

We have several main properties of the determinant $d(A)$. We can verify each of these properties based on the first or the second definition of the determinant. We often use the first definition to avoid repeating well-known arguments.

(**1**). $d(E) = 1$.

(**2**). *The determinant d is a polylinear function of rows of the matrix.*

Property (2) follows from the relations $d(A) = \eta A$ and the similar property of the ordinary determinant.   □

(**3**). *If a matrix A' is obtained from the matrix A by a permutation of the ith row and the jth row for $i \neq j$, then $d(A') = -s_{iji} d(A)$.*

Proof The ith row of the matrix A' is equal to $1_{ij} \circ A_j$ and the jth row is equal to $1_{ji} \circ A_i$. The ith row of the matrix $\eta(A')$ is equal to $(t_{i1} s_{ij1} a_{j1}, \ldots, t_{in} s_{ijn} a_{jn})$. For every k, we obtain that

$$t_{ik} s_{ijk} = s_{ik\ell} s_{ijk} = s_{ij\ell} s_{jk\ell} = s_{ij\ell} t_{jk}.$$

We can repeat this action for the jth row of the matrix $\eta(A')$. Then

$$|\eta(A')| = s_{ij\ell} s_{ji\ell} |A''| = s_{iji} |A''|,$$

where the matrix A'' is obtained from ηA by the permutation of the ith row and the jth row. Therefore, $|A''| = -|\eta A|$. Now we obtain

$$d(A') = |\eta(A')| = s_{iji} |A''| = -s_{iji} |\eta A| = -s_{iji} d(A).   \quad \square$$

(**4**). *If some two rows of the matrix A are proportional, then $d(A) = 0$.*

Proof Let the jth row of the matrix A be equal to $r_{ji} \circ A_i$ for some $r_{ji} \in R$. By property (2), we can assume that this row is of the form $(s_{ji1} a_{i1}, \ldots, s_{ijn} a_{in})$. For the matrix ηA, the ith row is equal to $(s_{i1\ell} a_{i1}, \ldots, s_{in\ell} a_{in})$ for some ℓ, and the jth is equal to $(s_{j1\ell} s_{ji1} a_{i1}, \ldots, s_{jn\ell} s_{jin} a_{in})$. For every $k = 1, \ldots, n$, we have $s_{jik} s_{jk\ell} = s_{ji\ell} s_{ik\ell}$. Now it is clear that the ith row of the matrix ηA is proportional to the jth row of ηA. Consequently, $d(A) = |\eta A| = 0.$ □

(5). *If we multiply (in the sense of the operation $\circ$) the jth row of the matrix A by some element $r_{ij} \in R$ and add the product to the ith row of the matrix A, then the determinant of the obtained matrix is equal to $d(A)$.*

The proof of (5) uses standard arguments based on properties (2) and (4). □

(6). *For any matrices A and B the relation $d(AB) = d(A)d(B)$ holds.*

Proof It is clear that

$$d(AB) = |\eta(AB)| = |\eta(A)\eta(B)| = |\eta(A)||\eta(B)| = d(A)d(B). \quad □$$

(7). *If $s_{ikj} = s_{jki}$ for all i, j, k, then $d(A) = d(A^t)$ for each matrix A. If none of the elements s_{iji} are zero-divisors in R, then the converse also holds.*

Proof We can use the homomorphism η or the complete development formula; in fact, they are is the same. We use the first method. We have the relations $d(A) = |\eta A|$ and $d(A^t) = |\eta(A^t)|$. There exists a known correspondence between the summands of determinants $|\eta A|$ and $|\eta(A^t)|$. We take some summand c of the determinant $|\eta A|$; let τ be a permutation of subscripts of this summand. We have $\tau = \sigma_1 \cdot \ldots \cdot \sigma_m$, where σ_i are pairwise independent cycles (in particular, of length 1). Further, let c_i be the product of multipliers whose subscripts belong to σ_i, $i = 1, \ldots, m$. Then $c = c_1 \cdot \ldots \cdot c_m$.

Let a summand d of the determinant $|\eta(A^t)|$ correspond to c. The permutation of its subscripts is $\tau^{-1} = \sigma_m^{-1} \cdot \ldots \cdot \sigma_1^{-1}$. We have the corresponding representation $d = d_1 \cdot \ldots \cdot d_m$ of the element d. We verify that $c_1 = d_1, \ldots, c_m = d_m$. With this purpose in mind, we take some cycle $\sigma = (i_1 i_2 \ldots i_k)$. Without loss of generality, we can assume that $k \geq 2$. Then $\sigma^{-1} = (i_k i_{k-1} \ldots i_1)$. Now it is sufficient to verify that the product $t_{i_1 i_2} t_{i_2 i_3} \cdot \ldots \cdot t_{i_{k-1} i_k} t_{i_k i_1}$ is equal to the product $t_{i_k i_{k-1}} t_{i_{k-1} i_{k-2}} \cdot \ldots \cdot t_{i_2 i_1} t_{i_1 i_k}$. These products are equal to

$$(t_{i_1 i_2} t_{i_2 i_3} \cdot \ldots \cdot t_{i_{k-2} i_{k-1}} t_{i_{k-1} i_1}) s_{i_{k-1} i_k i_1} \quad \text{and}$$
$$(t_{i_2 i_1} t_{i_3 i_2} \cdot \ldots \cdot t_{i_{k-1} i_{k-2}} t_{i_1 i_{k-1}}) s_{i_1 i_k i_{k-1}},$$

respectively. The form of the expressions in parentheses suggests that we can use induction on the length of the cycle σ. The case $k = 2$ is obvious.

Now we assume that all multipliers s_{iji} are not zero-divisors and $d(A) = d(A^t)$ for every matrix A. If some two of the three subscripts i, j, k coincide, then $s_{ikj} = s_{jki}$. Therefore, we assume that the subscripts i, j, k are pairwise distinct. We take the matrix

$$A = E + E_{ik} + E_{kj} + E_{ji} - E_{ii} - E_{kk} - E_{jj}.$$

The determinant $d(A)$ is equal to $1_{ik} \circ 1_{kj} \circ 1_{ji} = s_{ikj} s_{iji}$ and the determinant $d(A^t)$ is equal to $s_{jki} s_{jij}$. Consequently, $s_{ikj} = s_{jki}$. □

For a determinant with corner consisting of zeros, the known formula holds.

(8). If a matrix A is of the form $\begin{pmatrix} B & D \\ 0 & C \end{pmatrix}$, where B and C are matrices of orders m and $n - m$, respectively, then $d(A) = d(B)d(C)$.

Proof The matrix B has the multiplier system $\{s_{ijk} \mid 1 \le i, j, k \le m\}$ and the matrix C has the multiplier system $\{s_{ijk} \mid m + 1 \le i, j, k \le n\}$ (all subscripts can be decreased by m). For the rings $M(m, R, \{s_{ijk}\})$ and $M(n - m, R, \{s_{ijk}\})$, we assume that the corresponding homomorphisms η are restrictions of the homomorphism η for the ring $M(n, R, \{s_{ijk}\})$. It is understood that the matrix B is identified with the matrix $\begin{pmatrix} B & 0 \\ 0 & 0 \end{pmatrix}$, and the matrix C identified with the matrix $\begin{pmatrix} 0 & 0 \\ 0 & C \end{pmatrix}$. Now we have the relations

$$d(A) = |\eta A| = |\eta B| \cdot |\eta C| = d(B)d(C). \qquad \square$$

If there are zero divisors among the multipliers s_{ijk}, then this often complicates work with matrices. We define a formal matrix ring, which sometimes helps to avoid these difficulties. In one case, a similar ring is introduced in [106]; see below.

We fix a commutative ring R and an integer $n \ge 2$. Let $X = \{x_{ijk}\}$ be a set consisting of $n(n^2 - 1)$ variables, where $1 \le i, j, k \le n, i \ne j, j \ne k$. Further, let $R[X]$ be the polynomial ring in variables x_{ijk} with coefficients in the ring R.

Let I be the ideal of the ring $R[X]$ generated by all differences of the form $x_{ijk}x_{ik\ell} - x_{ij\ell}x_{jk\ell}$. We denote by $\overline{R[X]}$ the factor ring $R[X]/I$. We identify elements of the ring R with their images in $\overline{R[X]}$. For simplicity, we denote the residue class $x_{ijk} + I$ by x_{ijk}. In what follows, it is important that the elements x_{ijk} are not zero-divisors in $\overline{R[X]}$.

We assume that $x_{iik} = 1 = x_{ikk}$ for all $i, k = 1, \ldots, n$. In the ring $\overline{R[X]}$, the relations $x_{ijk}x_{ik\ell} = x_{ij\ell}x_{jk\ell}$ hold for all values of subscripts. Therefore, the formal matrix ring $M(n, \overline{R[X]}, \{x_{ijk}\})$ exists. We denote this ring by $M(n, \overline{R[X]}, X)$. There exists a known homomorphism

$$\eta \colon M(n, \overline{R[X]}, X) \to M(n, \overline{R[X]});$$

see the paragraph before Proposition 4.1.2. In addition, the determinant $d \colon M(n, \overline{R[X]}, X) \to \overline{R[X]}$ was defined earlier. In some sense, the ring $M(n, \overline{R[X]}, X)$ and the determinant d form a couniversal object for formal matrix rings of order n over R and their determinants. We assign a precise meaning to this phrase.

Assume that we have a concrete formal matrix ring $M(n, R, \{s_{ijk}\})$. In this case, we have several homomorphisms; we call them the *exchange homomorphisms* and denote them by the same symbol θ. Every such homomorphism replaces the symbol x_{ijk} by the element s_{ijk}. First of all, this is a homomorphism $\theta \colon R[X] \to R$. Since R is embedded in $R[X]$, we have that θ splits, $R[X] = R \oplus \mathrm{Ker}(\theta)$.

It follows from the main identities (4.1.1) of Sect. 4.1 that $I \subseteq \mathrm{Ker}\,\theta$. After identification, we can assume that $\overline{R[X]} = R \oplus (\mathrm{Ker}(\theta))/I$. In addition, θ induces the homomorphism $\overline{R[X]} \to R$; we denote it by the same symbol θ.

More generally, there exists a split exchange homomorphism $\theta: M(n, R[X]) \to M(n, R)$ such that θ is applied to each element of the matrix, and there also exists a decomposition $M(n, R[X]) = M(n, R) \oplus \mathrm{Ker}(\theta)$. Then θ induces the exchange homomorphism

$$\theta: M(n, \overline{R[X]}) \to M(n, R)$$

and the decomposition

$$M(n, \overline{R[X]}) = M(n, R) \oplus (\mathrm{Ker}(\theta)/M(n, I)). \tag{$*$}$$

The last homomorphism θ is also a homomorphism of formal matrix rings

$$M(n, \overline{R[X]}, X) \to M(n, R, \{s_{ijk}\}).$$

Remarks

(1). The decomposition $(*)$ is only additive.

(2). The homomorphism θ is surjective. More precisely, every matrix $(a_{ij}) \in M(n, R)$ is the image of the matrix (a_{ij}) from $M(n, \overline{R[X]})$ (we identify the matrices (a_{ij}) and $(a_{ij} + I)$).

Now we have two commutative diagrams:

$$M(n, \overline{R[X]}, X) \xrightarrow{\eta} M(n, \overline{R[X]})$$

$$\theta \downarrow \qquad\qquad\qquad \theta \downarrow$$

$$M(n, R, \{s_{ijk}\}) \xrightarrow{\eta} M(n, R) \quad,$$

$$M(n, \overline{R[X]}) \xrightarrow{\det} \overline{R[X]}$$

$$\theta \downarrow \qquad\qquad \theta \downarrow$$

$$M(n, R) \xrightarrow{\det} R \quad,$$

where det is an ordinary determinant. The commutativity of the second diagram is directly verified using the complete development formula of the determinant. We also have a commutative diagram which connects the previous two diagrams:

$$M(n, \overline{R[X]}, X) \xrightarrow{\ d\ } \overline{R[X]}$$

$$\theta \downarrow \qquad\qquad \theta \downarrow$$

$$M(n, R, \{s_{ijk}\}) \xrightarrow{\ d\ } R \ .$$

When we say that the pair $(M(n, \overline{R[X]}, X), d)$ is couniversal, we mean that the last diagram exists.

In one important case, we can simplify the construction of the ring $M(n, \overline{R[X]}, X)$. We consider formal matrix rings $M(n, R, \{s_{ijk}\})$ such that every multiplier s_{ijk} is a non-negative integral degree of some non-zero element s. We denote by $M(n, R, s)$ some such ring. In Sects. 4.3 and 4.4, we studied in detail rings $M(n, R, s)$ such that each multiplier s_{ijk} is equal to 1 or s. In Sect. 4.5, we assumed that $s_{ijk} = s^m, m \geq 1$, for $i \neq j$ and $j \neq k$.

We also assume that $s^k \neq s^\ell$ for any distinct non-negative k and ℓ. Let x be a variable. For any i, j, k, we set $x_{ijk} = x^m$ provided $s_{ijk} = s^m$. The set $\{x_{ijk} \,|\, i, j, k = 1, \ldots, n\}$ is a multiplier system in the ring of polynomials $R[x]$. Consequently, there exists a formal matrix ring $M(n, R[x], \{x_{ijk}\})$; we denote it by $M(n, R[x], x)$. The above three diagrams turn into the diagrams

$$M(n, R[x], x) \xrightarrow{\ \eta\ } M(n, R[x])$$

$$\theta \downarrow \qquad\qquad \theta \downarrow$$

$$M(n, R, s) \xrightarrow{\ \eta\ } M(n, R) \ ,$$

$$M(n, R[x], x) \xrightarrow{\ d\ } R[x]$$

$$\theta \downarrow \qquad\qquad \theta \downarrow$$

$$M(n, R, s) \xrightarrow{\ d\ } R \ .$$

In these diagrams, d is the determinant and θ is the exchange homomorphism which replaces the symbol x by the symbol s. Tang and Zhou define and use the ring $M(n, R[x], x)$ in [106].

4.7 Some Theorems About Formal Matrices

We preserve the notation of the previous section. As above, we assume that there is some formal matrix ring $M(n, R, \{s_{ijk}\})$, where R is a commutative ring. We show that there exist analogues of the Cayley–Hamilton theorem and one known invertibility criterion of a matrix.

We recall that $t_{ij} = s_{ij\ell}$ for some fixed ℓ. Throughout, we use the homomorphism

$$\eta\colon M(n, R, \{s_{ijk}\}) \to M(n, R), \quad (a_{ij}) \to (t_{ij}a_{ij});$$

see Sect. 4.1.

Let $A = (a_{ij})$ be a matrix. We denote by $\eta(A)^*$ the adjoint matrix for ηA. Then

$$\eta(A)\eta(A)^* = \eta(A)^*\eta(A) = |\eta(A)|E.$$

Then we write $\eta(A)^* = (A'_{ji})$, where A'_{ji} is the cofactor of the element $t_{ji}a_{ji}$. We recall that A'_{ji} is the determinant of the matrix which is obtained from the matrix ηA by replacing the element $t_{ji}a_{ji}$ by 1 and by replacing all the remaining elements of the jth row and the ith column by 0.

We temporarily assume that all elements s_{ijk} are non-zero-divisors. First, we also assume that $n \geq 3$. We take an arbitrary summand of the determinant A'_{ji}, where $j \neq i$. This summand necessarily contains the multiplier $t_{ik}a_{ik}t_{kj}a_{kj}$ for some $k \neq i, j$. Since $t_{ik}t_{kj} = s_{ikj}t_{ij}$, we have $A'_{ji} = t_{ji}A_{ji}$, where A_{ji} is a particular element of the ring R obtained from A'_{ji} by the canonical method. If $i = j$, $t_{ij} = 1$ and $A_{ji} = A'_{ji}$. For $n = 2$, the existence of such an element t_{ij} is directly verified.

We consider the matrix $A^* = (A_{ji})$. We have $\eta(A^*) = \eta(A)^*$. Then

$$\eta(AA^*) = \eta(A)\eta(A^*) = \eta(A)\eta(A)^* = |\eta(A)| \cdot E = d(A)E = \eta(d(A)E);$$

we similarly have

$$\eta(A^*A) = \eta(d(A)E).$$

Since all s_{ijk} are non-zero-divisors, $\mathrm{Ker}(\eta) = 0$ by Proposition 4.1.2. Consequently, $AA^* = A^*A = d(A)E$.

How do we construct the matrix A^* if not all elements s_{ijk} are non-zero-divisors? For this purpose, we use the commutative diagram from Sect. 4.6:

$$
\begin{array}{ccc}
M(n, \overline{R[X]}, X) & \xrightarrow{\ d_X\ } & \overline{R[X]} \\[2mm]
\theta \downarrow & & \theta \downarrow \\[2mm]
M(n, R, \{s_{ijk}\}) & \xrightarrow{\ d\ } & R
\end{array}
\tag{4.7.1}
$$

(here we have replaced d by d_X in the first row of the diagram). Since all x_{ijk} are non-zero-divisors, it follows from the above that there exists a uniquely determined matrix $A^*_X \in M(n, \overline{R[X]}, X)$ such that $AA^*_X = A^*_X A = d_X(A)E$. Therefore, we obtain that

$$\theta(A)\theta(A_X^*) = \theta(A_X^*)\theta(A) = \theta(d_X(A)E) \quad \text{or}$$
$$A\theta(A_X^*) = \theta(A_X^*)A = \theta(d_X(A)E).$$

It follows from the diagram that $\theta d_X(A) = d\theta(A) = d(A)$, whence $\theta(d_X(A)E) = d(A)E$. As a result, we obtain the relations

$$A\theta(A_X^*) = \theta(A_X^*)A = d(A)E.$$

It remains to set $A^* = \theta(A_X^*)$. If all elements s_{ijk} are non-zero-divisors, then this matrix A^* coincides with the matrix A^* defined in the text before the diagram. If we compare the similar methods of constructing the matrix A^* described before diagram (4.7.1) and the matrix A_X^*, then it is clear that $A^* = \theta(A_X^*)$.

We finish the first part of this section with the following results.

Theorem 4.7.1 *Assume that we have a formal matrix ring $M(n, R, \{s_{ijk}\})$ and let A be a matrix from this ring.*

(1) $AA^* = AA^* = d(A)E.$
(2) *The matrix A is invertible if and only if $d(A)$ is an invertible element of the ring A.*
(3) *If A is an invertible matrix, then $A^{-1} = d(A)^{-1}A^*$.*

For the determinant $d(A)$, there are analogous decompositions of the determinant with respect to elements of the row and the orthogonality property of rows and cofactors. We mean the relations

$$a_{i1} \circ A_{i1} + a_{i2} \circ A_{i2} + \ldots + a_{in} \circ A_{in} = d(A),$$
$$a_{i1} \circ A_{j1} + a_{i2} \circ A_{j2} + \ldots + a_{in} \circ A_{jn} = 0, \quad i \neq j$$

following from the relations $AA^* = d(A)E$.

By the Cayley–Hamilton theorem, every matrix is a root of its characteristic polynomial. We extend this theorem to matrices in the matrix ring $M(n, R, \{s_{ijk}\})$.

Let x be a variable. We have a formal matrix ring $M(n, R[x], \{s_{ijk}\})$ and the homomorphism $\eta\colon M(n, R[x], \{s_{ijk}\}) \to M(n, R[x])$ from Proposition 4.1.2. This homomorphism extends the homomorphism $\eta\colon M(n, R, \{s_{ijk}\}) \to M(n, R)$.

Now we take some symbol λ. Let A be a matrix in the ring $M(n, R, \{s_{ijk}\})$. Using the relations $f(\lambda) = d(\lambda E - A)$, we define the characteristic polynomial $f(\lambda)$ of the matrix A (with respect to the ring $M(n, R, \{s_{ijk}\})$), where $d(\lambda E - A)$ is the determinant of the matrix $\lambda E - A$ in the ring $M(n, R[\lambda], \{s_{ijk}\})$. Since $d(\lambda E - A) = |\eta(\lambda E - A)| = |\lambda E - \eta A|$, we have that $f(\lambda)$ is the characteristic polynomial of the matrix ηA in the ring $M(n, R)$. By the Hamilton-Cayley theorem, $f(\eta A) = 0$. With the use of the relation $f(\lambda) = \lambda^n + a_1\lambda^{n-1} + \ldots + a_{n-1}\lambda + a_n$, $a_i \in R$, we obtain that $\eta(f A) = f(\eta A) = 0$. If η is an injective mapping (all s_{ijk} are non-zero-divisors), then $f(A) = 0$; which is what we required.

If the mapping η is not injective, then we again consider the ring $M(n, \overline{R[X]}, X)$ from Sect. 4.6; the corresponding homomorphism η is injective for this ring. There exists a ring $M(n, (\overline{R[X]})[\lambda], X)$. The homomorphism $\theta: \overline{R[X]} \to R$ from Sect. 4.6 induces a homomorphism $\theta: (\overline{R[X]})[\lambda] \to R[\lambda]$, which applies θ to coefficients of polynomials. Thus, the last homomorphism θ induces the homomorphism

$$\theta: M(n, (\overline{R[X]})[\lambda], X) \to M(n, R[\lambda], \{s_{ijk}\}).$$

There also exists the following commutative diagram which is similar to the diagram from Sect. 4.6:

$$M(n, (\overline{R[X]})[\lambda], X) \xrightarrow{\;d\;} (\overline{R[X]})[\lambda]$$

$$\theta \downarrow \qquad\qquad\qquad \theta \downarrow \qquad\qquad\qquad (4.7.2)$$

$$M(n, R[\lambda], \{s_{ijk}\}) \xrightarrow{\;d\;} R[\lambda] \quad .$$

We return to the characteristic polynomial $f(\lambda)$ of the matrix $A = (a_{ij})$. We take the matrix $\overline{A} = (\overline{a}_{ij})$ from the ring $M(n, \overline{R[X]}, X)$ with the property $\theta(\overline{A}) = A$, where θ is taken from the diagram (4.7.1). Let $F(\lambda)$ be the characteristic polynomial of the matrix $\overline{A}$, $F(\lambda) = d(\lambda E - \overline{A})$, where $\lambda E - \overline{A} \in M(n, (\overline{R[X]})[\lambda], X)$. Then $\theta(\lambda E - \overline{A}) = \lambda E - A$, where θ is taken from diagram (4.7.2). It follows from the diagram (4.7.2) that

$$\theta d(\lambda E - \overline{A}) = d\theta(\lambda E - \overline{A}) = d(\lambda E - A).$$

enlargethispage11pt In other words, $\theta(F(\lambda)) = f(\lambda)$, where

$$F(\lambda) \in (\overline{R[X]})[\lambda], \quad f(\lambda) \in R[\lambda], \quad \theta: (\overline{R[X]})[\lambda] \to R[\lambda].$$

We represent the polynomial $F[\lambda]$ in more detail:

$$F(\lambda) = \lambda^n + \overline{a}_1 \lambda^{n-1} + \ldots + \overline{a}_{n-1}\lambda + \overline{a}_n, \quad \overline{a}_i \in \overline{R[X]}.$$

Then we calculate

$$f(\lambda) = \theta(F(\lambda)) = \lambda^n + a_1 \lambda^{n-1} + \ldots + a_{n-1}\lambda + a_n.$$

Now it follows from the above that

$$F(\overline{A}) = \overline{A}^n + \overline{a}_1 \overline{A}^{n-1} + \ldots + \overline{a}_{n-1}\overline{A} + \overline{a}_n E = 0.$$

Therefore,

$$0 = \theta(F(\overline{A})) = A^n + a_1 A^{n-1} + \ldots + a_{n-1}A + a_n E = f(A).$$

This proves the following.

Theorem 4.7.2 *If A is a matrix in the ring $M(n, R, \{s_{ijk}\})$ and $f(\lambda)$ is the characteristic polynomial of the matrix A, then $f(A) = 0$.*

Tang and Zhou [106] proved Theorem 4.7.2 for their formal matrix rings; see the beginning of Sect. 4.3.

Finally, we consider the uniqueness of the determinant. For the ordinary determinant of matrices, some combinations of properties (1)–(8) of Sect. 4.6 are characteristic properties. We have that if some mapping $f: M(n, R) \to R$ has some family of properties (1)–(8), then f coincides with the determinant. Under some restrictions to the determinant defined in Sect. 4.6, we show that properties (1)–(3) are characteristic in the specified sense.

Theorem 4.7.3 *Let $M(n, R, \{s_{ijk}\})$ be a formal matrix ring where the ring R does not have elements of additive order 2 and all multipliers s_{ijk} are not zero-divisors. In addition, let $f: M(n, R, \{s_{ijk}\}) \to R$ be a mapping with the following properties.*

(1) $f(E)=1$.
(2) f is a polylinear function of rows of the matrix.
(3) If we interchange the ith and jth rows in an arbitrary matrix $A \in M(n, R, \{s_{ijk}\})$ and denote the obtained matrix by A', then $f(A') = -s_{iji} f(A)$.

Then f coincides with the determinant d of the ring $M(n, R, \{s_{ijk}\})$.

Proof Let $A = (a_{ij})$ be an arbitrary matrix. We use polylinearity of the function f several times to represent the element $f(A)$ as a sum of elements of the form $f(C)$, where C is a matrix whose ith row of the matrix C contains the element a_{ij} in position (i, j) and the remaining elements are equal to zero. From each row of the matrix C, we take out this element a_{ij} from the sign of the mapping f. We denote by D the obtained matrix (each row of the matrix D contains at most one 1, and the remaining positions consist of zeros). By considering (4.7.2), we obtain that $f(C) = a_{1j_1} \cdot \ldots \cdot a_{njn} f(D)$.

We show that if there are equal subscripts among subscripts $j_1, \ldots, j_n$, then $f(D) = 0$. For example, let 1 be in positions (i, k) and (j, k), $i \neq j$. In the matrix D, we interchange the ith and jth rows; we denote the obtained matrix by D'. By (3) $f(D') = -s_{iji} f(D)$. It follows from (4.7.2) that

$$f(D') = s_{ijk}s_{jik} f(D) = s_{iji} f(D).$$

Therefore, we have the relations

$$s_{iji} f(D) = -s_{iji} f(D), \quad 2s_{iji} f(D) = 0, \quad 2f(D) = 0, \quad f(D) = 0.$$

Thus, there remain only elements in $f(C)$ such that the subscripts of elements $a_{1j_1}, \ldots, a_{nj_n}$ form a permutation. In addition, if some row of the matrix D consists of zeros, then $f(D) = 0$ by property (2). Therefore, we assume that each row of the matrix D contains precisely one non-zero element, which is equal to 1; this means that D is a permutation matrix. Let τ be a permutation of subscripts of non-zero elements of the matrix D. We have $\tau = \sigma_1 \cdot \ldots \cdot \sigma_m$, where σ_i are pairwise independent cycles. We denote one of these cycles by σ; let $\sigma = (i_1 i_2 \ldots i_k)$. By interchanging the rows $i_1, i_2, \ldots, i_k$, we obtain a matrix such that the units from these rows are on the main diagonal. What happens to the element $f(D)$ in this situation? We answer this via induction on the length of the cycle σ.

For $m = 2$ and $m = 3$, the induction hypothesis is directly verified. We assume that for any cycle σ of length m, where $3 \le m < k$, the relation

$$f(D) = (-1)^{m-1} s_{i_1 i_2 i_3} s_{i_1 i_3 i_4} \cdot \ldots \cdot s_{i_1 i_{m-1} i_m} s_{i_1 i_m i_1} f(V)$$

holds. The difference between matrices V and D is that the rows $i_1, \ldots, i_m$ of the matrix V have 1s on the main diagonal.

Now let σ be of length k. We interchange the rows i_1 and i_2 of the matrix D and denote the obtained matrix by D'. For the matrices D and D', we have the relation $f(D') = -s_{i_1 i_2 i_1} f(D)$. In the matrix D', we take out from the sign of f the element $s_{i_1 i_2 i_3}$ of the row i_1 and the element $s_{i_2 i_1 i_3}$ of the row i_2. We denote the obtained matrix by D''. We have the relation $f(D') = s_{i_2 i_1 i_2} s_{i_1 i_2 i_3} f(D'')$. It follows from the last two relations that $f(D) = -s_{i_1 i_2 i_3} f(D'')$. We have a 1 in position (i_2, i_2) of the matrix D'', and the subscripts of the elements (units) of the remaining rows form a cycle $(i_1 i_3 \ldots i_k)$ of length $k - 1$. By the induction hypothesis, we obtain the relation

$$f(D'') = (-1)^{k-2} s_{i_1 i_3 i_4} s_{i_1 i_4 i_5} \cdot \ldots \cdot s_{i_1 i_{k-1} i_k} s_{i_1 i_k i_1} f(W),$$

where W is a matrix which has 1s in rows $i_1, \ldots, i_k$ on the main diagonal and the remaining rows coincide with the corresponding rows of the matrix D. As a result, we obtain the relation

$$f(D) = (-1)^{k-1} s_{i_1 i_2 i_3} s_{i_1 i_3 i_4} \cdot \ldots \cdot s_{i_1 i_{k-1} i_k} s_{i_1 i_k i_1} f(W).$$

Since the function d of the determinant satisfies properties (1)–(3), a similar relation holds for d.

Next, we proceed similarly with the remaining cycles σ_i from the decomposition of the permutation τ. Ultimately, we obtain that the corresponding matrix W coincides with the identity matrix. Therefore, $f(W) = d(W)$. Finally, we obtain that

$$f(D) = d(D), \quad f(C) = d(C), \quad f(A) = d(A). \qquad \square$$

Remarks **(1)**. A matrix $A = (a_{ij})$ which is not invertible in the ring $M(n, R)$ can be invertible in the ring $M(n, R, \Sigma)$, where Σ is some multiplier system. We formulate the following problem: Characterize the matrices which are invertible in some ring $M(n, R, \Sigma)$. Since there exists a homomorphism

$$\eta \colon M(n, R, \Sigma) \to M(n, R)$$

from Sect. 4.1, we state this problem in another form. For which matrices A and multiplier systems $\Sigma = \{s_{ijk}\}$ is the matrix $A' = (s_{ij\ell}a_{ij})$ invertible in the ring $M(n, R)$ for some integer $\ell = 1, 2, \ldots, n$?

(2). We can define and study the permanent of the matrix A in the ring $M(n, R, \Sigma)$. It is worth noting that for various problems concerning the determinant and the permanent, there are possible variations associated with the existence of many multiplier systems Σ. For example, in the case of the well-known Polya problem concerning conversion of the permanent and the determinant, we can require that the permanent of the matrix A in the ring $M(n, R, \Sigma)$ coincides with the determinant of some other matrix in the ring $M(n, R, \Sigma')$ for some (in general, other) multiplier system Σ'.

(3). Let F be a field and $\Sigma = \{s_{ijk}\}$ some multiplier system $s_{ijk} \in F$. If all s_{ijk} are non-zero, then by Proposition 4.1.2, there exists an isomorphism $M(n, F, \Sigma) \cong M(n, F)$. This case is not very interesting. If some multipliers s_{ijk} are equal to zero, then properties of the ring $M(n, F, \Sigma)$ can be very different from properties of the ring $M(n, F)$. We highlight the special case where each multiplier s_{ijk} is equal to 1 or 0. For the corresponding rings $M(n, F, \Sigma)$, we can study problems (I)–(III) formulated in Sect. 4.1. In addition, we can consider the above questions (1) and (2).

Chapter 5
Grothendieck and Whitehead Groups of Formal Matrix Rings

We develop some calculation methods for the Grothendieck group K_0 and the Whitehead group K_1 of formal matrix rings making use of the K_0 and K_1 groups of the original rings.

In Sect. 5.1, we define some functors between categories of finitely generated projective K-modules and $(R \times S)$-modules. These functors induce some homomorphisms between the groups $K_i(K)$ and $K_i(R)$, $K_i(S)$, $i = 0, 1$ which are used to calculate the groups $K_0(K)$ and $K_1(K)$. In Sect. 5.1, under some restrictions, we also prove the equivalence of categories of finitely generated projective K-modules and finitely generated projective $(R \times S)$-modules.

Sections 5.2 and 5.3 are devoted to the K_0 group of the formal matrix ring of order 2.

In Sect. 5.4, we study the K_1 group of a formal matrix ring of order 2.

In Sect. 5.5, we briefly consider the K_0 and K_1 groups of a formal matrix ring of an arbitrary order.

5.1 Equivalence of Two Categories of Projective Modules

Let R be a ring. We denote by $P(R)$ the category of all finitely generated projective R-modules.

The Grothendieck group $K_0(R)$ of the ring R is the factor group F/H, where F is a free Abelian group whose basis consists of the classes of isomorphic modules in $P(R)$, and H is the subgroup of F which is generated by all elements of the form $(X \oplus Y) - (X) - (Y)$ for all $X, Y \in P(R)$, where (X) denotes the class of modules which are isomorphic to X. Let $[X] = (X) + H \in K_0(R)$. Then $[X \oplus Y] = [X] + [Y]$.

Let $GL(n, R)$ be the group of invertible matrices of order n over the ring R. We assume that the group $GL(n, R)$ is embedded in the group $GL(n + 1, R)$ via the monomorphism $A \to \begin{pmatrix} A & 0 \\ 0 & 1 \end{pmatrix}$. We denote by $GL(R)$ the union of the groups

© Springer International Publishing AG 2017
P. Krylov and A. Tuganbaev, *Formal Matrices*,
Algebra and Applications 23, DOI 10.1007/978-3-319-53907-2_5

$GL(n, R)$ for all $n \geq 1$. Then $E(n, R)$ denotes the subgroup of $GL(n, R)$ generated by all elementary matrices. Under the embedding $GL(n, R)$ in $GL(n + 1, R)$, the subgroup $E(n, R)$ is embedded in $E(n + 1, R)$. The union of all $E(n, R)$ is denoted by $E(R)$ and is called the *group of elementary matrices*. This group coincides with the commutant of the group $GL(R)$. The *Whitehead group* $K_1(R)$ of the ring R is the Abelian group $GL(R)/E(R)$.

For our purposes, another definition of the group $K_1(R)$ is more useful. We take the category of automorphisms in $P(R)$. The objects of the category are pairs of the form (X, α), where $X \in P(R)$ and α is an automorphism of the module X. The morphisms are defined in the usual way for such categories. Let F be a free Abelian group whose generator elements are pairs (X, α), one for each class of pairs which are isomorphic in the category of automorphisms. The group $K_1(R)$ is the factor group F/G, where the subgroup G is generated by elements of the form $(X, \alpha\beta) - (X, \alpha) - (X, \beta)$ and $(X \oplus Y, \alpha\beta) - (X, \alpha) - (Y, \beta)$; it is assumed that α acts identically on Y and β acts identically on X. The residue class $(X, \alpha) + G$ is denoted by $[(X, \alpha)]$.

A natural isomorphism between two groups $K_1(R)$ is proved as follows; see Theorem 1.2 of Chap. 9 in [12] and Theorem 3.1.7 in [97]. A matrix $A \in GL(n, R)$ determines an automorphism α of the module R^n. The element $[(R^n, \alpha)]$ of the group F/G corresponds to the element $AE(R)$ of the group $GL(R)/E(R)$. In this book, we use only the second definition of the group $K_1(R)$.

Every element of the group $K_1(R)$ can be represented in the form $[(X, \alpha)]$ for some X and α. For each X, we have that $[(X, 1)]$ is the zero element. The opposite element to $[(X, \alpha)]$ is $[(X, \alpha^{-1})]$. The automorphism α contains the main information about the element $[(X, \alpha)]$. Therefore, we often write $[\alpha]$ instead of $[(X, \alpha)]$.

For the definition of homomorphisms between the K_0 and K_1 groups, we use the following facts.

If R and S are two rings, then every additive functor $F \colon P(R) \to P(S)$ induces the group homomorphism

$$K_0(F) \colon K_0(R) \to K_0(S), \qquad [X] \to [FX]$$

(this is the action of the homomorphism on generator elements of the group $K_0(R)$) and

$$K_1(F) \colon K_1(R) \to K_1(S), \qquad [(X, \alpha)] \to [(FX, F\alpha)].$$

In particular, the ring homomorphism $i \colon R \to S$ induces the functor

$$T(i) = S \otimes_R (-) \colon P(R) \to P(S), \qquad X \to S \otimes_R X.$$

Consequently, we have the homomorphisms

$$K_0(i) \colon K_0(R) \to K_0(S), \quad [X] \to [S \otimes_R X], \quad \text{and}$$
$$K_1(i) \colon K_1(R) \to K_1(S), \quad [(X, \alpha)] \to [(S \otimes_R X, 1 \otimes \alpha)].$$

Now let $K = \begin{pmatrix} R & M \\ N & S \end{pmatrix}$ be the formal matrix ring with the trace ideals I and J defined in Sect. 2.1. There exists a commutative diagram of rings

$$
\begin{array}{ccc}
R \times S & \xrightarrow{k} & R/I \times S/J \\
i \downarrow & & \| \\
K & \xrightarrow{j} & R/I \times S/J
\end{array}
\tag{5.1.1}
$$

in which i is the diagonal embedding, $(r, s) \rightarrow \begin{pmatrix} r & 0 \\ 0 & s \end{pmatrix}$ and k is the canonical epimorphism, $j : \begin{pmatrix} r & m \\ n & s \end{pmatrix} \rightarrow (r+I, s+J)$. Then we can represent diagram functors

$$
\begin{array}{ccc}
P(R \times S) & \xrightarrow{T(k)} & P(R/I \times S/J) \\
T(i) \downarrow & & \| \\
P(K) & \xrightarrow{T(j)} & P(R/I \times S/J).
\end{array}
\tag{5.1.2}
$$

These functors act as follows. If $X \oplus Y$ is a projective $(R \times S)$-module, then $T(i)(X \oplus Y) = (X, T(X)) \oplus (T(Y), Y)$, i.e. $T(i)$ is the restriction of the functor T from Sect. 3.1, and

$$
T(k)(X \oplus Y) = (R/I \otimes_R X) \oplus (S/J \otimes_S Y) \cong X/IX \oplus Y/JY.
$$

If (X, Y) is a projective K-module, then $T(j)(X, Y) = X/MY \oplus Y/NX$, see the remark before Corollary 3.6.2. All three functors transfer homomorphisms to the induced homomorphisms. We can verify that the functors $T(j)T(i)$ and $T(k)$ are naturally equivalent.

Diagrams (5.1.1) and (5.1.2) induce a commutative diagram of the K_0 groups and their homomorphisms

$$
\begin{array}{ccc}
K_0(R) \oplus K_0(S) & \xrightarrow{K_0(k)} & K_0(R/I) \oplus K_0(S/J) \\
K_0(i) \downarrow & & \| \\
K_0(K) & \xrightarrow{K_0(j)} & K_0(R/I) \oplus K_0(S/J).
\end{array}
\tag{5.1.3}
$$

The homomorphisms act as follows (we define the homomorphisms only on generator elements of the K_0 groups):

$$
\begin{array}{l}
K_0(i): [X] + [Y] \rightarrow [(X, T(X))] + [(T(Y), Y)]. \quad \text{Then} \\
K_0(k): [X] + [Y] \rightarrow [X/IX] + [Y/JY]. \quad \text{Thus,} \\
K_0(j): [(X, Y)] \rightarrow [X/MY] + [Y/NX].
\end{array}
$$

Diagrams (5.1.1) and (5.1.2) also induce the following commutative diagram of the K_1 groups and their homomorphisms:

$$
\begin{array}{ccc}
K_1(R) \oplus K_1(S) & \xrightarrow{K_1(k)} & K_1(R/I) \oplus K_1(S/J) \\
K_1(i)\downarrow & & \parallel \\
K_1(K) & \xrightarrow{K_1(j)} & K_1(R/I) \oplus K_1(S/J).
\end{array} \qquad (5.1.4)
$$

For this diagram, we have that

$$
K_1(i)\colon [\alpha] + [\beta] \to [(\alpha, 1 \otimes \alpha)] + [(1 \otimes \beta, \beta)].
$$

Then we can represent

$$
K_1(k)\colon [\alpha] + [\beta] \to [\overline{\alpha}] + [\overline{\beta}],
$$

where $\overline{\alpha}$ (resp., $\overline{\beta}$) is an automorphism, induced by the automorphism α (resp., β) on X/IX (resp., Y/JY). Also, $K_1(j)\colon [(\alpha, \beta)] \to [\overline{\alpha}] + [\overline{\beta}]$, where $\overline{\alpha}$ (resp., $\overline{\beta}$) is the automorphism induced by the automorphism α (resp., β) on X/MY (resp., Y/NX).

We formulate a result about equivalence categories of projective modules. If L is an ideal of some ring T, then the set of all powers of the ideal L forms a basis of neighborhoods of zero of the so-called L-*adic* topology of the ring T.

Theorem 5.1.1 *Let $I \subseteq J(R)$, $J \subseteq J(S)$, the ring R complete in the I-adic topology and the ring S complete in the J-adic topology. Then the functor $T(i)$ defines an equivalence between the categories $P(R \times S)$ and $P(K)$.*

Proof By Theorem 2.4.1, the kernel $\begin{pmatrix} I & M \\ N & J \end{pmatrix}$ of the homomorphism j is contained in the radical $J(K)$. Consequently, the functor $T(K)$ acts bijectively, and the functor $T(j)$ acts injectively on the corresponding classes of isomorphic finitely generated projective modules; see Proposition 2.12 of Chap. 3s from [12].

Let (P, Q) be some finitely generated projective K-module. There exists a finitely generated projective $(R \times S)$-module $X \oplus Y$ such that $T(k)(X \oplus Y) \cong P/MQ \oplus Q/NP$. We obtain

$$
T(i)(X \oplus Y) \cong (P, Q) \quad \text{or} \quad (P, Q) \cong (X, T(X)) \oplus (T(Y), Y).
$$

We take arbitrary finitely generated projective R-modules X_1, X_2 and arbitrary finitely generated projective S-modules Y_1, Y_2. The mappings

$$
\begin{aligned}
T(i)\colon & \operatorname{Hom}_R(X_1, X_2) \to \operatorname{Hom}_K((X_1, T(X_1)), (X_2, T(X_2))), \\
T(i)\colon & \operatorname{Hom}_S(Y_1, Y_2) \to \operatorname{Hom}_K((T(Y_1), Y_1), (T(Y_2), Y_2))
\end{aligned}
$$

are isomorphisms; see Lemma 3.1.2. This proves that $T(i)$ is an equivalence.   □

In the proof of Theorem 5.1.1, we proved the following property (cf. Theorem 3.7.3).

Corollary 5.1.2 *Under the conditions of Theorem 5.1.1, for any finitely generated projective K-module (P, Q), there are a finitely generated projective R-module X and a finitely generated projective S-module Y such that $(P, Q) \cong (X, T(X)) \oplus (T(Y), Y)$.*

Corollary 5.1.3 *Under the conditions of Theorem 5.1.1, there are isomorphisms $K_i(K) \cong K_i(R) \oplus K_i(S)$, $i = 0, 1$.*

Proof In the case $i = 0$, the assertion follows from the equivalence of categories $P(R \times S)$ and $P(K)$. In the case $i = 1$, the assertion follows from the equivalence of the categories of automorphisms of categories $P(R \times S)$ and $P(K)$. $\square$

Remark 5.1.4 Any ring K with nilpotent trace ideals I and J satisfies the conditions of Theorem 5.1.1. For example, this is the case if K is a ring of formal triangular matrices, where $M = 0$ or $N = 0$.

5.2 The Group $K_0(A, B)$

We consider categories which are more general than the category $P(R)$ of finitely generated projective modules. Let A be an R-module. The direct summands of direct sums of a finite number of copies of the module A are called *finitely A-projective* R-modules [73, Sect. 32]. We denote by $P(A)$ the category of finitely A-projective R-modules. The categories $P(A)$ and $P(\text{End}_R A)$ are equivalent [73, Theorem 32.1].

The category $P(A)$ is additive and its classes of isomorphic objects form a set. Consequently, we can define the *Grothendieck group* $K_0(P(A))$ of the category $P(A)$ such that $K_0(R) = K_0(P(R))$ [12, 97]. By the above equivalence, we have a group isomorphism $K_0(P(A)) \cong K_0(\text{End}_R A)$. We use the symbol $K_0(A)$ instead of $K_0(P(A))$.

Similar to Sect. 5.1, we remark that the group $K_0(A)$ is the factor group F/H, where F is a free Abelian group such that the basis of F consists of classes of isomorphic modules from $P(A)$, and H is the subgroup of F generated by all elements of the form $(X \oplus Y) - (X) - (Y)$ for all $X, Y \in P(A)$, where (X) denotes the class of modules which are isomorphic to X. Let $[X] = (X) + H \in K_0(A)$. Then $[X \oplus Y] = [X] + [Y]$. If $z \in K_0(A)$, then $z = [X] - [Y]$ for some X and Y from $P(A)$. We use this useful fact without further mention. In addition, we note that the relation $[X] = [Y]$ in $K_0(A)$ is equivalent to the property that $X \oplus A^n \cong Y \oplus A^n$ for some positive integers n. In this case, we say that X and Y are *stably isomorphic* in $P(A)$. In particular, the modules X and Y from $P(R)$ are stably isomorphic if $X \oplus R^n \cong Y \oplus R^n$ for some positive integers n.

If A and C are two modules, then each additive functor $F: P(A) \to P(C)$ induces the group homomorphism $K_0(A) \to K_0(C)$, $[X] \to [FX]$ (we define the action of this homomorphism only on generator elements of the group $K_0(A)$).

We remark that it is not necessary to distinguish the categories $P(A)$ and $P(A^n)$, $n \geq 2$. Consequently, $K_0(A) \cong K_0(A^n)$.

Lemma 5.2.1 *Let A and B be two modules and the module A finitely B-projective. Then there exists an embedding of categories $P(A) \to P(B)$ which induces the homomorphism $K_0(A) \to K_0(B)$.*

Proof The assertion follows from the property that every finitely A-projective module is a finitely B-projective module.   □

Assume that we have a direct module decomposition $A = B \oplus C$. Since B and C are finitely A-projective modules, we have the homomorphisms $K_0(B) \to K_0(A)$ and $K_0(C) \to K_0(A)$ from Lemma 5.2.1.

Proposition 5.2.2 *If $A = B \oplus C$, where C is a finitely B-projective module, then there exists a canonical isomorphism $K_0(A) \cong K_0(B)$.*

Proof It is clear that A is a finitely B-projective module. We have the homomorphism $K_0(A) \to K_0(B)$ from Lemma 5.2.1. It is inverse to the homomorphism $K_0(B) \to K_0(A)$ specified before the proposition.   □

As earlier, let $K = \begin{pmatrix} R & M \\ N & S \end{pmatrix}$ be a formal matrix ring.

Up to the end of this section, we assume that (A, B) is some K-module, $P(A)$ and $P(B)$ are the above-defined categories and $P(A, B)$ is the category of finitely (A, B)-projective K-modules. Then $K_0(A)$ and $K_0(B)$ are the above-defined groups, and $K_0(A, B)$ is a similar group for the category $P(A, B)$.

Our purpose is to describe the group $K_0(A, B)$ using the groups $K_0(A)$ and $K_0(B)$. Here are two simple general results.

Corollary 5.2.3 *For any R-module X and every S-module Y, there exist isomorphisms $K_0(X, T(X)) \cong K_0(X)$ and $K_0(T(Y), Y) \cong K_0(Y)$.*

Proof By Corollary 3.1.4, there exist canonical ring isomorphisms $\operatorname{End}_K(X, T(X)) \cong \operatorname{End}_R X$ and $\operatorname{End}_K(T(Y), Y) \cong \operatorname{End}_S Y$. They induce the equivalence of categories $P(X, T(X))$ and $P(X)$, the equivalence of categories $P(T(Y), Y)$ and $P(Y)$, and isomorphisms of the corresponding K_0 groups.   □

Corollary 5.2.4 *For the ring K, let us have an equivalence situation, i.e. $I = R$ and $J = S$; see Sect. 3.8. Then $K_0(A) \cong K_0(A, B) \cong K_0(B)$.*

Proof By Corollary 3.8.2, there exist module isomorphisms $(A, T(A)) \cong (A, B) \cong (T(B), B)$. Now we apply Corollary 5.2.3.   □

We introduce brief notations for elements of the group $K_0(A, B)$. Let $z = [(X_1, Y_1)] - [(X_2, Y_2)] \in K_0(A, B)$. Set $x = [X_1] - [X_2] \in K_0(A)$ and $y = [Y_1] - [Y_2] \in K_0(B)$. We write the element z as the pair $[x, y]$. The relations

$$-[x, y] = [-x, -y], \quad [x, y] + [x', y'] = [x + x', y + y']$$

hold. Now we define some functors and homomorphisms between the categories defined above and the K_0 groups.

If $(X, Y) \in P(A, B)$, then $X \in P(A)$ and $Y \in P(B)$. We can define a functor $F: P(A, B) \to P(A)$ assuming that $F(X, Y) = X$ and homomorphisms are transformed to the induced homomorphisms. We also denote by the symbol F a similar functor $P(A, B) \to P(B)$, $(X, Y) \to Y$. These functors are restrictions of the functors $(1, 0)$ and $(0, 1)$ from Sect. 3.1. The functors F induce the homomorphisms

$$\mu_1: K_0(A, B) \to K_0(A), \quad [x, y] \to x,$$
$$\mu_2: K_0(A, B) \to K_0(B), \quad [x, y] \to y.$$

We also set

$$\mu = \mu_1 + \mu_2: K_0(A, B) \to K_0(A) \oplus K_0(B), \quad [x, y] \to x + y.$$

It is clear that

$$\mathrm{Ker}\,\mu = \mathrm{Ker}\,\mu_1 \cap \mathrm{Ker}\,\mu_2 \quad \text{and}$$
$$x = 0 = y \Leftrightarrow [x, y] \in \mathrm{Ker}\,\mu.$$

We make one assumption. It gives the possibility to define a homomorphism from the K_0 group in the "opposite" direction (compared with μ_1). Then we assume that $(A, T(A))$ is a finitely (A, B)-projective K-module (we recall that the K-module (A, B) is fixed). In such a case, for every finitely A-projective R-module X, the K-module $(X, T(X))$ is finitely (A, B)-projective.

We define the functor $P(A) \to P(A, B)$, $X \to (X, T(X))$ (homomorphisms are transformed to the induced homomorphisms, see Sect. 3.1). This functor is the restriction of the functor $(1, T_N)$ from Sect. 3.1.

The defined functor induces the homomorphism

$$\alpha_1: K_0(A) \to K_0(A, B),$$
$$[X_1] - [X_2] \to [(X_1, T(X_1))] - [(X_2, T(X_2))].$$

Let

$$\sigma = \alpha_1 \mu_2: K_0(A) \to K_0(B),$$
$$[X_1] - [X_2] \to [T(X_1)] - [T(X_2)].$$

If we set $x = [X_1] - [X_2]$, then it follows from the rule of representation of elements of the group $K_0(A, B)$ that we have the relation $\alpha_1(x) = [x, \sigma(x)]$. Since $\alpha_1 \mu_1 = 1$, we obtain the following corollary.

Corollary 5.2.5 *There exists a direct decomposition $K_0(A, B) = \mathrm{Im}\,\alpha_1 \oplus \mathrm{Ker}\,\mu_1$, where $\mathrm{Im}\,\alpha_1 \cong K_0(A)$.*

The representation of the element $[x, y]$ with respect to the direct sum from Corollary 5.2.5 has the form $[x, y] = [x, \sigma(x)] + [0, y - \sigma(x)]$. We remark that the correspondence $\alpha_1 : x \rightarrow [x, \sigma(x)]$ is an isomorphism from $K_0(A)$ onto Im α_1. We define the subgroup G of the group $K_0(B)$ to be $G = \{y - \sigma(x) \mid [x, y] \in K_0(A, B)\}$. Now we can find the structure of the factor group $K_0(A, B)/\operatorname{Ker} \mu$.

Theorem 5.2.6 *There exists an isomorphism*

$$K_0(A, B)/\operatorname{Ker} \mu \cong K_0(A) \oplus G, \quad [x, y] + \operatorname{Ker} \mu \rightarrow x + (y - \sigma(x)).$$

Proof The kernel of the homomorphism

$$K_0(A, B) \rightarrow K_0(A) \oplus G, \quad [x, y] \rightarrow x + (y - \sigma(x)),$$

consists of the elements $[x, y]$ with $x = 0 = y$. Thus, this kernel is equal to Ker μ. □

Apparently, it is not easy to find the structure of the subgroup Ker μ. It follows from Sect. 5.5 that this subgroup can be non-zero. If z is a non-zero element in Ker μ and $z = [(X_1, Y_1)] - [(X_2, Y_2)]$, then X_1, X_2 are stably isomorphic in $P(A)$ and Y_1, Y_2 are stably isomorphic in $P(B)$, but (X_1, Y_1) and (X_2, Y_2) are not stably isomorphic in $P(A, B)$.

Corollary 5.2.7 *The following assertions are equivalent.*

(a) $G = 0$.
(b) $y = \sigma(x)$ *for any* $[x, y] \in K_0(A, B)$.
(c) $K_0(A, B) = \operatorname{Im} \alpha_1 \oplus \operatorname{Ker} \mu$.

Proof The equivalence of assertions (a) and (b) follows from the definition of the subgroup G.

(b) $\Rightarrow$ (c). In our case, Ker $\mu_1 \subseteq$ Ker μ_2 and Ker $\mu =$ Ker μ_1. Now we can use Corollary 5.2.5.

(c) $\Rightarrow$ (a). Again, we obtain Ker $\mu_1 \subseteq$ Ker μ_2. Let $[x, y] \in K_0(A, B)$. Then

$$[x, y] - [x, \sigma(x)] = [0, y - \sigma(x)] \in \operatorname{Ker} \mu_1.$$

Consequently, $[0, y - \sigma(x)] \in \operatorname{Ker} \mu_2$, $y - \sigma(x) = 0$ and $G = 0$. □

In Corollary 5.2.5, Theorem 5.2.6, and Corollary 5.2.7, we assume that $(A, T(A))$ is a finitely (A, B)-projective K-module. Now let $(T(B), B)$ be a finitely (A, B)-projective K-module. The analogous assertions to the above are also true. We have the functor $P(B) \rightarrow P(A, B)$, $Y \rightarrow (T(Y), Y)$, and the induced homomorphism

$$\alpha_2 : K_0(B) \rightarrow K_0(A, B), \quad [Y_1] - [Y_2] \rightarrow [(T(Y_1), Y_1)] - [(T(Y_2), Y_2)].$$

Let

$$\tau = \alpha_2 \mu_1 : K_0(B) \rightarrow K_0(A), \quad [Y_1] - [Y_2] \rightarrow [T(Y_1)] - [T(Y_2)].$$

Then $\alpha_2(y) = [\tau(y), y]$, where $y = [Y_1] - [Y_2]$. In addition, the relation $\alpha_2 \mu_2 = 1$ holds and analogues of Corollary 5.2.5, Theorem 5.2.6 and Corollary 5.2.7 hold. Namely, there exists a direct decomposition

$$K_0(A, B) = \text{Im} \, \alpha_2 \oplus \text{Ker} \, \mu_2, \quad \text{where} \quad \text{Im} \, \alpha_2 \cong K_0(B).$$

Then there exists an isomorphism

$$K_0(A, B)/ \text{Ker} \, \mu \cong H \oplus K_0(B), \quad [x, y] + \text{Ker} \, \mu \to (x - \tau(y)) + y,$$

where $H = \{x - \tau(y) \,|\, [x, y] \in K_0(A, B)\}$ is a subgroup of the group $K_0(A)$.

Now we assume that both K-modules $(A, T(A))$ and $(T(B), B)$ are finitely (A, B)-projective. In this case, $\sigma\tau$ and $\tau\sigma$ are endomorphisms of the groups $K_0(A)$ and $K_0(B)$, respectively. We also set

$$\alpha = \alpha_1 + \alpha_2 \colon K_0(A) \oplus K_0(B) \to K_0(A, B).$$

The composition $\alpha\mu$ is an endomorphism of the group $K_0(A) \oplus K_0(B)$,

$$\alpha\mu(x + y) = (x + \tau(y)) + (\sigma(x) + y), \quad x \in K_0(A), \; y \in K_0(B).$$

With respect to this direct sum, this endomorphism can be represented by the matrix $\begin{pmatrix} 1 & \tau \\ \sigma & 1 \end{pmatrix}$.

Corollary 5.2.8 *If* $\begin{pmatrix} 1 & \tau \\ \sigma & 1 \end{pmatrix}$ *is an invertible matrix, then we have the relation* $K_0(A, B) = \text{Im} \, \alpha_1 \oplus \text{Im} \, \alpha_2 \oplus \text{Ker} \, \mu$ *and the isomorphism* $K_0(A, B) \cong K_0(A) \oplus K_0(B) \oplus \text{Ker} \, \mu$.

Proof In our case, $\alpha\mu$ is an automorphism of the group $K_0(A) \oplus K_0(B)$. Consequently, $\alpha(\mu\xi) = 1$ for some automorphism ξ of this group. Therefore,

$$K_0(A, B) = \text{Im} \, \alpha \oplus \text{Ker} \, \mu\xi = \text{Im} \, \alpha_1 \oplus \text{Im} \, \alpha_2 \oplus \text{Ker} \, \mu. \qquad \square$$

The invertibility of the endomorphism $1 - \sigma\tau$ of the group $K_0(A)$ is equivalent to the invertibility of the endomorphism $1 - \tau\sigma$ of the group $K_0(B)$. In this case, $\begin{pmatrix} 1 & \sigma \\ \tau & 1 \end{pmatrix}$ is an invertible matrix with inverse matrix $\begin{pmatrix} 1 & -\sigma \\ -\tau & 1 \end{pmatrix}$.

Corollary 5.2.9 *If* $1 - \sigma\tau$ *is an automorphism of the group* $K_0(A)$ *(for example, this is the case if the endomorphism* $\sigma\tau$ *is nilpotent or is contained in the radical* $J(\text{End} \, K_0(A))$*), then the assertion of Corollary 5.2.8 holds for the group* $K_0(A, B)$.

The following result complements Corollary 5.2.7.

Corollary 5.2.10 *If $G = 0 = H$, then σ and τ are mutually inverse isomorphisms and*

$$K_0(A, B) \cong K_0(A) \oplus \operatorname{Ker} \mu \cong K_0(B) \oplus \operatorname{Ker} \mu.$$

Remark 5.2.11 Finally we remark that the relations

$$G = (1 - \tau\sigma)K_0(B) + \sigma(H) \quad \text{and}$$
$$H = (1 - \sigma\tau)K_0(A) + \tau(G)$$

both hold.

5.3 K_0 Groups of Formal Matrix Rings

We apply the results of Sect. 5.2 to the formal matrix ring $\begin{pmatrix} R & M \\ N & S \end{pmatrix}$. Thus, we take the K-module K as the module (A, B). We can represent this module in the form $((R, M), (N, S))$ or $(R \oplus M, N \oplus S)$. In several cases, we calculate the group $K_0(K)$ using the groups $K_0(R)$ and $K_0(S)$ (sometimes up to the subgroup $\operatorname{Ker} \mu$).

First, we present two specific results related to extreme situations. The first result follows from Corollary 5.1.3.

Corollary 5.3.1 *If the trace ideals I and J of the ring K are nilpotent, then $K_0(K) \cong K_0(R) \oplus K_0(S)$.*

If the homomorphism $\psi \colon N \otimes_R M \to S$ from Sect. 2.1 is an epimorphism, then we say that K is a *semi-surjective Morita context*.

Corollary 5.3.2 *(1) Reference [12] If K is a semi-surjective Morita context, then*
   $K_0(K) \cong K_0(R)$.
(2) If K is an equivalence situation, i.e. $I = R$ and $J = S$, then $K_0(R) \cong K_0(K) \cong K_0(S)$ (cf. Corollary 5.2.4).

Proof **(1)**. We have

$$T(R \oplus M) \cong (N \otimes_R R) \oplus (N \otimes_R M) \cong N \oplus (N \otimes_R M).$$

The mapping ψ is an isomorphism, see Sect. 3.8. Consequently,

$$T(R \oplus M) \cong N \oplus S \quad \text{and} \quad K \cong (R \oplus M, T(R \oplus M)).$$

By Corollary 5.2.3, $K_0(K) \cong K_0(R \oplus M)$. The R-module M is a finitely generated projective R-module, see Sect. 3.8. It follows from Proposition 5.2.2 that $K_0(R \oplus M) \cong K_0(R)$. Thus, $K_0(K) \cong K_0(R)$.

   (2). The assertion is a special case of (1). $\qquad\qquad\qquad\qquad\qquad\qquad\qquad\square$

We take functors $T_N = N \otimes_R (-): R\text{-Mod} \to S\text{-mod}$, $T_M = M \otimes_S (-):$ $S\text{-Mod} \to R\text{-Mod}$ from Sect. 3.1. We usually denote them by the same symbol T. The restrictions of these functors provide functors $T: P(R) \to P(N)$ and $T: P(S) \to P(M)$. The functors T induce the homomorphisms

$$\sigma: K_0(R) \to K_0(N), \quad [X_1] - [X_2] \to [T(X_1)] - [T(X_2)], \quad \text{and}$$
$$\tau: K_0(S) \to K_0(M), \quad [Y_1] - [Y_2] \to [T(Y_1)] - [T(Y_2)].$$

The functors $(1, T_N): R\text{-Mod} \to K\text{-Mod}$ and $(T_M, 1): S\text{-Mod} \to K\text{-Mod}$ from Sect. 3.1 provide functors

$$P(R) \to P(K), \quad X \to (X, T(X)), \quad \text{and}$$
$$P(S) \to P(K), \quad Y \to (T(Y), Y)$$

which induce the homomorphisms

$$\alpha_1: K_0(R) \to K_0(K), \quad \alpha_2: K_0(S) \to K_0(K).$$

Using the notation of Sect. 5.2 for elements of the group $K_0(K)$, we have

$$\alpha_1(x) = [x, \sigma(x)], \quad x \in K_0(R), \quad \text{and} \quad \alpha_2(y) = [\tau(y), y], \quad y \in K_0(S).$$

Now we set $\alpha = \alpha_1 + \alpha_2$.

The homomorphisms σ, τ, α_1, α_2 are similar to the corresponding homomorphisms from Sect. 5.2.

We also have two functors

$$F: P(K) \to P(R \oplus M), \quad F: P(K) \to P(N \oplus S)$$

and the corresponding homomorphisms

$$\mu_1: K_0(K) \to K_0(R \oplus M), \quad \mu_2: K_0(K) \to K_0(N \oplus S), \quad \mu = \mu_1 + \mu_2;$$

see the text after Corollary 5.2.4.

For the formal matrix ring K, we have diagram (5.1.3) from Sect. 5.1. We formulate it again in the form

$$
\begin{array}{ccc}
K_0(R) \oplus K_0(S) & \xrightarrow{\pi} & K_0(R/I) \oplus K_0(S/J) \\
\alpha \downarrow & & \| \\
K_0(K) & \xrightarrow{\rho} & K_0(R/I) \oplus K_0(S/J),
\end{array}
$$

where $\alpha = K_0(i)$, $\pi = K_0(k)$ and $\rho = K_0(j)$. We represent the action of these homomorphisms in a more compact form. Set

$$x = [X_1] - [X_2] \in K_0(R), \quad y = [Y_1] - [Y_2] \in K_0(S).$$

Then $\alpha(x + y) = [x, \sigma(x)] + [\tau(y), y]$. We denote $[X_1/IX_1] - [X_2/IX_2]$ and $[Y_1/JY_1]-[Y_2/JY_2]$ by x/Ix and y/Jy, respectively. Then $\pi(x+y) = x/Ix+y/Jy$. If $[x, y] \in K_0(K)$, where $x = [X_1] - [X_2]$ and $y = [Y_1] - [Y_2]$, then let

$$x/My = [X_1/MY_1] - [X_2/MY_2], \quad y/Nx = [Y_1/NX_1] - [Y_2/NX_2].$$

Then $\rho([x, y]) = x/My + y/Nx$.

By Corollary 5.1.3, we have that $K_0(K) \cong K_0(R) \oplus K_0(S)$ if the trace ideals I and J of the ring K are nilpotent. In addition, the homomorphism $K_0(i)$ from diagram (5.1.3) of Sect. 5.1 is an isomorphism; consequently, the homomorphism α is an isomorphism. We describe below how these isomorphisms act.

Now consider the more general situation where $I \subseteq J(R)$ and $J \subseteq J(S)$. First, we recall a well-known fact.

Let T be some ring, L an ideal of the ring T contained in $J(T)$ and $e: T \to T/L$ the canonical epimorphism. Then the kernel of the induced homomorphism $K_0(e): K_0(T) \to K_0(T/L)$ is equal to the zero subgroup. (In addition, if we assume that the ring T is complete in the L-adic topology, then $K_0(e)$ is an isomorphism by Proposition 1.3 of Chap. 9 in [12].)

We return to the formal matrix ring K. If $I \subseteq J(R)$ and $J \subseteq J(S)$, then the kernel $\begin{pmatrix} I & M \\ N & J \end{pmatrix}$ of the homomorphism j from diagram (5.1.1) in Sect. 5.1 is contained in the radical $J(K)$ (this was already used in the proof of Theorem 5.1.1). Consequently, π, ρ and α are monomorphisms. After natural identifications, we can assume that

$$K_0(R) \oplus K_0(S) \subseteq K_0(K) \subseteq K_0(R/I) \oplus K_0(S/J).$$

Theorem 5.3.3 *We assume that $I \subseteq J(R)$ and $J \subseteq J(S)$. The relation $K_0(K) = K_0(R) \oplus K_0(S)$ holds if and only if $\mathrm{Im}\,\rho = \mathrm{Im}\,\pi$.*

Theorem 5.3.3 follows from the property that the relation $\mathrm{Im}\,\rho = \mathrm{Im}\,\pi$ holds if and only if α is an isomorphism. $\qquad\square$

How do the isomorphisms α and α^{-1} act? (See also Corollary 5.1.3.) We take the element $[x, y] \in K_0(K)$. Then

$$x/My = x'/Ix' \quad \text{and} \quad y/Nx = y'/Jy', \quad \text{where} \quad x' \in K_0(R), y' \in K_0(S),$$

and x', y' are unique. The correspondence $[x, y] \to x' + y'$ defines the isomorphism α^{-1}, and the correspondence $x' + y' \to [x, y]$ defines the isomorphism α. Then we obtain that $K_0(K) = \mathrm{Im}\,\alpha_1 \oplus \mathrm{Im}\,\alpha_2$ and $[x, y] = [x', \sigma(x')] + [\tau(y'), y']$ is the representation of the element $[x, y]$ with respect to this direct sum.

In Sect. 5.2, we assumed that at least one of the K-modules $(A, T(A))$, $(T(B), B)$ is finitely (A, B)-projective. We assume that $_RM$ is a finitely generated projective module. Then $(R \oplus M, T(R \oplus M))$ is a finitely generated projective K-module by Proposition 3.7.1, i.e. the specified assumption holds for the K-module K. In this case, if $_SY$ is a finitely generated projective module, then $T(Y) = M \otimes_S Y$ is a finitely generated projective R-module. In addition, if (X, Y) is a finitely generated projective K-module, then X is a finitely generated projective R-module.

In such a situation, we can find the functors T, F and the corresponding homomorphisms τ and μ_1 defined in Sect. 5.2. Indeed, we obtain a "new" functor $T: P(S) \rightarrow P(R)$, $Y \rightarrow T(Y)$, and a "new" homomorphism $\tau: K_0(S) \rightarrow K_0(R)$ instead of $T: P(S) \rightarrow P(M)$ and $\tau: K_0(S) \rightarrow K_0(M)$, respectively. Then we have a "new" functor $F: P(K) \rightarrow P(R)$, $(X, Y) \rightarrow X$, and a "new" homomorphism $\mu_1: K_0(K) \rightarrow K_0(R)$, $[x, y] \rightarrow x$ instead of $F: P(K) \rightarrow P(R \oplus M)$ and $\mu_1: K_0(K) \rightarrow K_0(R \oplus M)$, respectively.

Since $\alpha_1 \mu_1 = 1$, Corollary 5.3.4 holds.

Corollary 5.3.4 *If M is a finitely generated projective R-module, then there exists a direct decomposition*

$$K_0(K) = \operatorname{Im} \alpha_1 \oplus \operatorname{Ker} \mu_1, \quad where \quad \operatorname{Im} \alpha_1 \cong K_0(R).$$

If $_S N$ is a finitely generated projective module, then we can similarly obtain "new" functors $T: P(R) \rightarrow P(S)$ and $F: P(K) \rightarrow P(S)$, $(X, Y) \rightarrow Y$, and "new" homomorphisms

$$\sigma: K_0(R) \rightarrow K_0(S), \quad [X_1] - [X_2] \rightarrow [T(X_1)] - [T(X_2)],$$
$$\mu_2: K_0(K) \rightarrow K_0(S), \quad [x, y] \rightarrow y.$$

In addition, $\alpha_2 \mu_2 = 1$. Consequently, we have the direct decomposition $K_0(K) = \operatorname{Im} \alpha_2 \oplus \operatorname{Ker} \mu_2$, where $\operatorname{Im} \alpha_2 \cong K_0(S)$.

Now we assume that $_R M$, $_S N$ are both finitely generated projective modules. In this case, the analogues of Theorem 5.2.6 and Corollaries 5.2.7–5.2.10 hold provided we set $\mu = \mu_1 + \mu_2$ and replace $K_0(A, B)$, $K_0(A)$ and $K_0(B)$ by $K_0(K)$, $K_0(R)$ and $K_0(S)$, respectively. We formulate only the analogue of Theorem 5.2.6. First, we define the subgroup $G = \{y - \sigma(x) \mid [x, y] \in K_0(K)\}$ of the group $K_0(S)$ and the subgroup $H = \{x - \tau(y) \mid [x, y] \in K_0(K)\}$ of the group $K_0(R)$.

Theorem 5.3.5 *We assume that $_R M$ and $_S N$ are finitely generated projective modules. There exist isomorphisms*

$$K_0(K)/\operatorname{Ker} \mu \cong K_0(R) \oplus G, \quad [x, y] + \operatorname{Ker} \mu \rightarrow x + (y - \sigma(x)),$$
$$K_0(K)/\operatorname{Ker} \mu \cong H \oplus K_0(S), \quad [x, y] + \operatorname{Ker} \mu \rightarrow (x - \tau(y)) + y.$$

Here is one case where the assumption of the analogue of Corollary 5.2.9 holds. If the formal matrix ring $K = \begin{pmatrix} R & M \\ N & S \end{pmatrix}$ is given and the homomorphism $\varphi: M \otimes_S N \rightarrow R$ is a monomorphism (see Sect. 2.1), then we say that K is a *semi-injective Morita context* (cf. Corollary 5.3.2(1)).

Corollary 5.3.6 *Let the conditions of Theorem 5.3.5 hold, K a semi-injective Morita context and the trace ideal I nilpotent. Then there exists an isomorphism*

$$K_0(K) \cong K_0(R) \oplus K_0(S) \oplus \operatorname{Ker} \mu.$$

Proof Let $x = [X_1] - [X_2] \in K_0(R)$. Then $\sigma\tau(x) = [M \otimes_S N \otimes_R X_1] - [M \otimes_S N \otimes_R X_2]$. Since $M \otimes_S N \cong I$ and the projective R-modules X_1, X_2 are flat, we obtain the canonical isomorphisms $M \otimes_S N \otimes_R X_1 \cong I X_1$, $M \otimes_S N \otimes_R X_2 \cong I X_2$ and the relation $\sigma\tau(x) = [I X_1] - [I X_2]$, $x \in K_0(R)$. We denote the last difference by $I x$. Thus, we have the relation $\sigma\tau(x) = I x, x \in K_0(R)$. Then we use induction on n to obtain that $(\sigma\tau)^n(x) = I^n x$ for any $n > 1$, where $I^n x$ denotes $[I^n X_1] - [I^n X_2]$. It is clear that $(\sigma\tau)^k = 0$ for some k. It remains to apply Corollary 5.2.9. $\square$

As above, we assume that $_R M$ and $_S N$ are finitely generated projective modules. We return to the functor (T, T) from Sect. 3.1. We recall that (T, T) is the functor K-Mod $\to K$-Mod such that

$$(X, Y) \to (T(Y), T(X)), \quad (\alpha, \beta) \to (T(\beta), T(\alpha)).$$

The K-module $(T, T)(K)$ is a finitely generated projective module. Therefore, for any finitely generated projective K-module (A, B), we have that $(T(B), T(A))$ is a finitely generated projective K-module. Consequently, we have the additive functor $(T, T)\colon P(K) \to P(K)$. We denote by $[\sigma, \tau]$ the endomorphism of the group $K_0(K)$ induced by the functor (T, T). Thus, $[\sigma, \tau]([x, y]) = [\tau(y), \sigma(x)]$.

5.4   The K_1 Group of a Formal Matrix Ring

Similar to the previous section, in this section we consider a formal matrix ring $K = \begin{pmatrix} R & M \\ N & S \end{pmatrix}$. Under some restrictions, we describe the Whitehead group $K_1(K)$ using the groups $K_1(R)$ and $K_1(S)$. The corresponding relations sometimes contain some group $\operatorname{Ker} \pi$ with unclear structure. It is interesting that the results and arguments of this section are similar to the corresponding results and arguments from Sect. 5.3.

We assume that $_R M$ and $_S N$ are finitely generated projective modules. The functors $T\colon P(R) \to P(S)$ and $T\colon P(S) \to P(R)$ induce the homomorphisms $e\colon K_1(R) \to K_1(S)$ and $h\colon K_1(S) \to K_1(R)$ where $e\colon [\alpha] \to [1 \otimes \alpha]$ and $h\colon [\beta] \to [1 \otimes \beta]$; see Sects. 5.1 and 5.3. Consequently, eh is an endomorphism of the group $K_1(R)$ and he is an endomorphism of the group $K_1(S)$.

We denote by θ_1 the restriction of the homomorphism $K_1(i)$ from diagram (5.1.4) in Sect. 5.1 to the group $K_1(R)$; the restriction of $K_1(i)$ to $K_1(S)$ is denoted by θ_2. Thus,

$$\theta_1: K_1(R) \to K_1(K), \quad [\alpha] \to [(\alpha, 1 \otimes \alpha)],$$
$$\theta_2: K_1(S) \to K_1(K), \quad [\beta] \to [(1 \otimes \beta, \beta)].$$

The functors $F: P(K) \to P(R)$ and $F: P(K) \to P(S)$ from Sect. 5.3 lead to the homomorphisms

$$\pi_1: K_1(K) \to K_1(R), \quad [(\alpha, \beta)] \to [\alpha],$$
$$\pi_2: K_1(K) \to K_1(S), \quad [(\alpha, \beta)] \to [\beta].$$

Set $\pi = \pi_1 + \pi_2$. It is clear that $\text{Ker}\,\pi = \text{Ker}\,\pi_1 \cap \text{Ker}\,\pi_2$.

We have the relations $\theta_1 \pi_1 = 1$ and $\theta_2 \pi_2 = 1$.

If we only have that $_R M$ is a finitely generated projective module, then π_1 still exists and $\theta_1 \pi_1 = 1$. Therefore, we have the following assertion.

Corollary 5.4.1 *If $_R M$ is a finitely generated projective module, then there exists a direct decomposition*

$$K_1(K) = \text{Im}\,\theta_1 \oplus \text{Ker}\,\pi_1, \quad \text{where} \quad \text{Im}\,\theta_1 \cong K_1(R).$$

With respect to this direct sum, the element $[(\alpha, \beta)]$ can be written in the form

$$[(\alpha, \beta)] = [(\alpha, 1 \otimes \alpha)] + [(1, \beta(1 \otimes \alpha)^{-1})].$$

Up to the end of this section, we assume that $_R M$ and $_S N$ are both finitely generated projective modules. Then we also have

$$K_1(K) = \text{Im}\,\theta_2 \oplus \text{Ker}\,\pi_2, \quad \text{where} \quad \text{Im}\,\theta_2 \cong K_1(S).$$

For the group $K_1(K)$, analogues of Theorems 5.2.6 and 5.3.5 hold. These theorems involve the subgroups G and H. We shall define the corresponding subgroups of the group $K_1(K)$ in a more meaningful way.

We consider existing direct decompositions

$$K_1(K) = \text{Im}\,\theta_1 \oplus \text{Ker}\,\pi_1 = \text{Im}\,\theta_2 \oplus \text{Ker}\,\pi_2.$$

The kernel of the restriction of the projection $K_1(K) \to \text{Im}\,\theta_2$ to $\text{Ker}\,\pi_1$ is equal to $\text{Ker}\,\pi$. Let B' be the image of this restriction. We denote by B the image of the subgroup B' under the isomorphism

$$\text{Im}\,\theta_2 \to K_1(S), \quad [(1 \otimes \beta, \beta)] \to [\beta].$$

Thus, B is a subgroup of $K_1(S)$. We describe the form of its elements. The representation of the element $[(\alpha, \beta)]$ of the group $K_1(K)$ with respect to the first decomposition is pointed out above. The representation of this element with respect to the second decomposition is

$$[(\alpha, \beta)] = [(1 \otimes \beta, \beta)] + [(\alpha(1 \otimes \beta)^{-1}, 1)].$$

Now it is clear that $B = \{[\beta(1 \otimes \alpha)^{-1}] \mid [(\alpha, \beta)] \in K_1(K)\}$. Similarly, we define the subgroup A of the group $K_1(R)$, by assuming $A = \{[\alpha(1 \otimes \beta)^{-1}] \mid [(\alpha, \beta)] \in K_1(K)\}$. We remark that

$$[\beta(1 \otimes \alpha)^{-1}] = [\beta] - e([\alpha]), \quad [\alpha(1 \otimes \beta)^{-1}] = [\alpha] - h([\beta]).$$

We formulate a general result.

Theorem 5.4.2 *There exist isomorphisms*

$$K_1(K)/\operatorname{Ker} \pi \cong K_1(R) \oplus B, \quad [(\alpha, \beta)] + \operatorname{Ker} \pi \to [\alpha] + [\beta(1 \otimes \alpha)^{-1}],$$
$$K_1(K)/\operatorname{Ker} \pi \cong A \oplus K_1(S), \quad [(\alpha, \beta)] + \operatorname{Ker} \pi \to [\alpha(1 \otimes \beta)^{-1}] + [\beta],$$

where B and A are the above-defined subgroups of $K_1(S)$ and $K_1(R)$.

Concerning the subgroup $\operatorname{Ker} \pi$, a question arises: Can the subgroup $\operatorname{Ker} \pi$ be non-zero? The answer is contained in Sect. 5.5.

We can precisely specify some parts of the groups B and A.

Lemma 5.4.3 *The following relations hold*

$$(1 - he)K_1(S) \subseteq B, \quad (1 - eh)K_1(R) \subseteq A.$$

Proof For the element $[\beta] \in K_1(S)$, we have

$$(1 - he)([\beta]) = [\beta] - [1 \otimes 1 \otimes \beta] = [\beta(1 \otimes 1 \otimes \beta)^{-1}].$$

Since $[(1 \otimes \beta, \beta)] \in K_1(K)$, we have $[\beta(1 \otimes 1 \otimes \beta)^{-1}] \in B$. The second inclusion is similarly verified.   □

In one quite clear case, Theorem 5.4.2 can be clarified.

Corollary 5.4.4 *If $1 - eh$ is an automorphism of the group $K_1(R)$ (for example, this is the case if the endomorphism eh is nilpotent or $eh \in J(\operatorname{End} K_1(R))$), then we have the relation $K_1(K) = \operatorname{Im} \theta_1 \oplus \operatorname{Im} \theta_2 \oplus \operatorname{Ker} \pi$ and an isomorphism $K_1(K) \cong K_1(R) \oplus K_1(S) \oplus \operatorname{Ker} \pi$.*

Proof Set $\theta = \theta_1 + \theta_2$. The composition $\theta\pi$ is an endomorphism of the group $K_1(R) \oplus K_1(S)$,

$$[\alpha] + [\beta] \to ([\alpha] + [1 \otimes \beta]) + ([1 \otimes \alpha] + [\beta])$$
$$= ([\alpha] + h([\beta])) + (e([\alpha]) + [\beta]).$$

With respect to this direct sum, the endomorphism $\theta\pi$ is represented by the matrix $\begin{pmatrix} 1 & e \\ h & 1 \end{pmatrix}$. The invertibility of the element $1 - eh$ of the ring $\mathrm{End}\, K_1(R)$ is equivalent to the invertibility of the element $1 - he$ of the ring $\mathrm{End}\, K_1(S)$. In this case, $\begin{pmatrix} 1 & e \\ h & 1 \end{pmatrix}$ is an invertible matrix with inverse matrix $\begin{pmatrix} 1 & -e \\ -h & 1 \end{pmatrix}$. We obtain that $\theta\pi$ is an automorphism of the group $K_1(R) \oplus K_1(S)$. Therefore, $\theta(\pi w) = 1$ for some automorphism w of the group $K_1(R) \oplus K_1(S)$. Therefore,

$$K_1(K) = \mathrm{Im}\,\theta \oplus \mathrm{Ker}\,\pi w = \mathrm{Im}\,\theta_1 \oplus \mathrm{Im}\,\theta_2 \oplus \mathrm{Ker}\,\pi. \qquad \square$$

Here is a case where the assumptions of Corollary 5.4.4 hold.

Corollary 5.4.5 *Let K be a semi-injective Morita context (see Corollary 5.3.6). If I is a nilpotent ideal, then*

$$K_1(K) \cong K_1(R) \oplus K_1(S) \oplus \mathrm{Ker}\,\pi.$$

Proof We have $(eh)([(X, \alpha)]) = [(M \otimes_S N \otimes_R X, 1 \otimes 1 \otimes \alpha)]$. Since $M \otimes_S N \cong I$ and the projective R-module X is flat, we have the canonical isomorphism $M \otimes_S N \otimes_R X \cong IX$. Then we obtain the relation $[(M \otimes_S N \otimes_R X, 1 \otimes 1 \otimes \alpha)] = [(IX, \overline{\alpha})]$, where $\overline{\alpha}$ is the restriction of α to IX. Then we obtain that $(eh)^n([(X, \alpha)]) = [(I^n X, \overline{\alpha})]$ for $n > 1$. It is clear that $(eh)^k = 0$ for some k. It remains to use Corollary 5.4.4. $\square$

We point out one important special case of Theorem 5.4.2, where one of the subgroups B or A is equal to the zero subgroup.

Corollary 5.4.6 **(1)** *The following assertions are equivalent:*

(a) $B = 0$;
(b) $[\beta] = [1 \otimes \alpha] = e([\alpha])$ *for any element* $[(\alpha, \beta)] \in K_1(K)$;
(c) $K_1(K) = \mathrm{Im}\,\theta_1 \oplus \mathrm{Ker}\,\pi \cong K_1(R) \oplus \mathrm{Ker}\,\pi.$

(2) If $B = 0 = A$, then e and h are mutually inverse isomorphisms and $K_1(K) \cong K_1(R) \oplus \mathrm{Ker}\,\pi \cong K_1(S) \oplus \mathrm{Ker}\,\pi$.

Proof **(1)**. The equivalence (a) $\Leftrightarrow$ (b) follows from the definition of the subgroup B.
(b) $\Rightarrow$ (c). We have $B = 0$. Consequently, $\mathrm{Ker}\,\pi_1 \subseteq \mathrm{Ker}\,\pi_2$ and $K_1(K) = \mathrm{Im}\,\theta_1 \oplus \mathrm{Ker}\,\pi$, where $\mathrm{Im}\,\theta_1 \cong K_1(R)$.
(c) $\Rightarrow$ (a). Again, we obtain $\mathrm{Ker}\,\pi_1 \subseteq \mathrm{Ker}\,\pi_2$, whence $B = 0$.
(2). The assertion follows from (1). $\square$

Corollary 5.4.7 (cf. Corollary 5.3.2) *If the ring K is an equivalence situation, then $K_1(R) \cong K_1(K) \cong K_1(S)$.*

Proof In our case, the assertion (b) of Corollary 5.4.6 holds. In addition, $\mathrm{Ker}\,\pi = 0$. It is also sufficient to specify that the categories $P(K)$, $P(R)$ and $P(S)$ are pairwise equivalent, see Theorem 3.8.5. $\square$

Remark 5.4.8 Similar to the case of the K_0 group (see the end of Sect. 2.2(5), the functor (T, T) induces an endomorphism of the group $K_1(K)$. We denote this endomorphism by $[e, h]$. We have the relation $[e, h]([(\alpha, \beta)]) = [(1 \otimes \beta, 1 \otimes \alpha)]$, $[(\alpha, \beta)] \in K_1(K)$.

5.5 K_0 and K_1 Groups of Matrix Rings of Order $n \geq 2$

In this section, we obtain some information about K_0 and K_1 groups of formal matrix rings of arbitrary order n. We follow similar arguments to the case $n = 2$. Sometimes, we can use the results for the case $n = 2$ by representing formal matrix rings of order $n > 2$ in the form of rings of formal block matrices of order 2.

Let K be a formal matrix ring

$$\begin{pmatrix} R_1 & M_{12} & \ldots & M_{1n} \\ M_{21} & R_2 & \ldots & M_{2n} \\ \ldots & \ldots & \ldots & \ldots \\ M_{n1} & M_{n2} & \ldots & R_n \end{pmatrix}$$

of order n with trace ideals $I_1, \ldots, I_n$ (see Sect. 2.3). We have the following commutative diagram of rings

$$\begin{array}{ccc} R_1 \times \ldots \times R_n & \xrightarrow{k} & R_1/I_1 \times \ldots \times R_n/I_n \\ i \downarrow & & \| \\ K & \xrightarrow{j} & R_1/I_1 \times \ldots \times R_n/I_n, \end{array} \qquad (5.5.1)$$

where i is the diagonal embedding, k is the canonical epimorphism, and

$$j: \begin{pmatrix} r_1 & a_{12} & \ldots & a_{1n} \\ a_{21} & r_2 & \ldots & a_{2n} \\ \ldots & \ldots & \ldots & \ldots \\ a_{n1} & a_{n2} & \ldots & r_n \end{pmatrix} \rightarrow (r_1 + I_1, \ldots, r_n + I_n).$$

Then we have the following diagram of functors

$$\begin{array}{ccc} P(R_1 \times \ldots \times R_n) & \xrightarrow{T(k)} & P(R_1/I_1 \times \ldots \times R_n/I_n) \\ T(i) \downarrow & & \| \\ P(K) & \xrightarrow{T(j)} & P(R_1/I_1 \times \ldots \times R_n/I_n). \end{array} \qquad (5.5.2)$$

The action of functors of the form $T(i)$ is described in Sect. 5.1.

Diagram (5.5.2) induces the commutative diagram of the K_0 groups and their homomorphisms

$$K_0(R_1) \oplus \ldots \oplus K_0(R_n) \xrightarrow{K_0(k)} K_0(R_1/I_1) \oplus \ldots \oplus K_0(R_n/I_n)$$
$$K_0(i) \downarrow \qquad\qquad\qquad\qquad \| \qquad\qquad\qquad (5.5.3)$$
$$K_0(K) \qquad \xrightarrow{K_0(j)} K_0(R_1/I_1) \oplus \ldots \oplus K_0(R_n/I_n)$$

and the same diagram, where the subscript 0 is replaced by the subscript 1. The action of the homomorphisms of the form $K_0(i)$, $K_1(i)$, and so on, is given in Sect. 5.1. We recall that the end of Sect. 3.1 contains some information about the structure and properties of modules over the formal matrix ring of order n.

Theorem 5.5.1 *Let K be a formal matrix ring of order n and $I_1, \ldots, I_n$ the trace ideals of K. If $I_k \subseteq J(R_k)$ and the ring R_k is complete in the I_k-adic topology for every $k = 1, \ldots, n$, then*

(1) the functor $T(i)$ is an equivalence;
(2) there exists a group isomorphism

$$K_i(K) \cong K_i(R_1) \oplus \ldots \oplus K_i(R_n), \quad i = 0, 1.$$

Proof **(1)**. The kernel of the homomorphism j is equal to

$$\begin{pmatrix} I_1 & M_{12} & \ldots & M_{1n} \\ M_{21} & I_2 & \ldots & M_{2n} \\ \ldots & \ldots & \ldots & \ldots \\ M_{n1} & M_{n2} & \ldots & I_n \end{pmatrix};$$

by Theorem 2.4.3, it is contained in the radical $J(K)$. On the corresponding classes of isomorphic finitely generated projective modules, the functor $T(k)$ acts bijectively and the functor $T(j)$ acts injectively.

Then it is possible to essentially repeat the argument from the proof of Theorem 5.1.1.

(2). The assertion directly follows from (1).   □

A ring K with nilpotent trace ideals satisfies the conditions of Theorem 5.5.1. We only assume that there are inclusions $I_1 \subseteq J(R_1), \ldots, I_n \subseteq J(R_n)$. Similar to the case $n = 2$ (see the paragraph before Theorem 5.3.3), we can assume after identification that we have the relations

$$K_0(R_1) \oplus \ldots \oplus K_0(R_n) \subseteq K_0(K) \subseteq K_0(R_1/I_1) \oplus \ldots \oplus K_0(R_n/I_n).$$

We apply Theorem 5.5.1 to formal matrix rings over the ring R, which are defined in Sect. 4.1. Let $K = M(n, R, \Sigma)$ be the ring of formal matrices of order $n \geq 2$ over the ring R where $\Sigma = \{s_{ijk} \mid i, j, k = 1, \ldots, n\}$ is a multiplier system of the ring K.

First of all, we remark that if all multipliers s_{ijk} are invertible elements, then $K_0(K) \cong K_0(R)$ and $K_1(K) \cong K_1(R)$. Indeed, it follows from Proposition 4.1.2 that there is an isomorphism $K \cong M(n, R)$. Consequently, $K_i(K) \cong K_i(M(n, R)) \cong K_i(R)$ (see Corollaries 5.3.2 and 5.4.7).

We can precisely specify the trace ideals of the ring K. Namely, the relation $I_k = \sum_{i \neq k} s_{kik} R$ holds for each $k = 1, \ldots, n$. We now formulate a corollary of Theorem 5.5.1 for an arbitrary ring K. We take a ring K such that each multiplier $s_{iji}, i \neq j$, is equal to s^m for some $m \geq 1$, where s is some central element of the ring R. Examples of such rings K are given in Sect. 4.4. We denote by $M(n, R, s)$ any such a ring.

Corollary 5.5.2 *Let K be the above-defined ring $M(n, R, s)$.*

(1) If s is an invertible element, then $K_i(M(n, R, s)) \cong K_i(R), i = 0, 1$.
(2) If s is a nilpotent element, then $K_i(M(n, R, s)) \cong K_i(R)^n, i = 0, 1$.
(3) If $s \in J(R)$, then

$$K_0(R)^n \subseteq K_0(K) \subseteq K_0(R/sR)^n.$$

Let us consider in more detail the case of the formal matrix ring $M(2, R, s)$ of order 2 over the ring R. The initial information about such rings and their modules is given at the end of Sect. 4.1. First, we formulate a special case of Corollary 5.5.2.

Corollary 5.5.3 *For the ring $K = M(2, R, s)$, assertions (1)–(3) of Corollary 5.5.2 hold for $n = 2$.*

In what follows, we assume that K is a ring of the form $M(2, R, s)$. We obtain some information about the groups $K_0(K)$ and $K_1(K)$.

We begin with the group $K_0(K)$. The homomorphisms σ and τ of the K_0 groups, defined in Sect. 5.3, are identity automorphisms in the considered situation. Defined at the end of Sect. 5.3, the endomorphism $[\sigma, \tau]$ obtains the form $[1, 1]$; it acts with the use of the relation $[x, y] \to [y, x], [x, y] \in K_0(K)$. Consequently, $[1, 1]$ is an involution. It isomorphically maps $\operatorname{Im} \alpha_1$ onto $\operatorname{Im} \alpha_2$ and $\operatorname{Ker} \mu_1$ onto $\operatorname{Ker} \mu_2$ (see Sects. 5.2 and 5.3 for more about the homomorphisms α_i and μ_i).

We take the homomorphisms $\alpha_1, \alpha_2 : K_0(R) \to K_0(K)$, where $\alpha_1(x) = [x, \sigma(x)]$ and $\alpha_2(x) = [\tau(x), x], x \in K_0(R)$ (it is possible to set $[x, x]' = \alpha_1(x)$ and $[x, x]'' = \alpha_2(x)$). Now we set $d = \alpha_1 - \alpha_2$ and $D = \operatorname{Im} d$. We have the isomorphism $K_0(R)/\operatorname{Ker} d \cong D$, where $\operatorname{Ker} d = \{x \in K_0(R) \mid \alpha_1(x) = \alpha_2(x)\}$. It is important for us that $D \subseteq \operatorname{Ker} \mu$.

We assume that the multiplier s is contained in the radical $J(R)$. Then the mapping $\alpha = (\alpha_1, \alpha_2) : K_0(R) \oplus K_0(R) \to K_0(K)$ is a monomorphism (see the paragraph before Theorem 5.3.3). It is clear that d is also a monomorphism and $D \cong K_0(R)$. In particular, $\operatorname{Ker} \mu \neq 0$ (see the text before Corollary 5.2.7).

As above, let $s \in J(R)$; in addition, we assume that the ring R is complete in the sR-adic topology (this is the case if s is a nilpotent element). Under these assumptions, we have the relations $K_0(K) = \operatorname{Im} \alpha_1 \oplus \operatorname{Im} \alpha_2 \cong K_0(R) \oplus K_0(R)$; see Corollaries 5.1.3, 5.3.1 and Theorem 5.3.3. It is clear that the relations $K_0(K) = \operatorname{Im} \alpha_i \oplus D, i = 1, 2$, also hold. It follows from the relations $D \subseteq \operatorname{Ker} \mu = \operatorname{Ker} \mu_1 \cap \operatorname{Ker} \mu_2$ and $K_0(K) = \operatorname{Im} \alpha_i \oplus \operatorname{Ker} \mu_i, i = 1, 2$, (see Corollaries 5.2.5 and 5.3.4) that $D = \operatorname{Ker} \mu$. By applying Corollary 5.2.7, we also obtain that $G = 0 = H$; see Sects. 5.2 and 5.3 for more about the subgroups G and H.

We pass to the group $K_1(K)$. The endomorphism $[e, h]$ of the group $K_1(K)$ from Sect. 5.4 has the form $[1, 1]$. Thus, $[1, 1]: [(\alpha, \beta)] \rightarrow [(\beta, \alpha)]$, $[(\alpha, \beta)] \in K_1(K)$. Consequently, $[1, 1]$ is an involution which isomorphically maps $\operatorname{Im} \theta_1$ and $\operatorname{Ker} \pi_1$ onto $\operatorname{Im} \theta_2$ and $\operatorname{Ker} \pi_2$, respectively; the homomorphisms θ_i and π_i, $i = 1, 2$, are defined in Sect. 5.4. In this situation, θ_1 and θ_2 are the homomorphisms $K_1(R) \rightarrow K_1(K)$ such that $\theta_1([\alpha]) = [(\alpha, 1 \otimes \alpha)]$ and $\theta_2([\alpha]) = [(1 \otimes \alpha, \alpha)]$, $[\alpha] \in K_1(R)$. We have more explicit relations

$$\theta_1([(X, \alpha)]) = [((X, T(X)), (\alpha, 1 \otimes \alpha))] \quad \text{and}$$
$$\theta_2([(X, \alpha)]) = [((T(X), X), (1 \otimes \alpha, \alpha))].$$

By considering the identifications and notations of Sects. 5.1 and 5.4, we obtain the relations

$$\theta_1([\alpha]) = [((X, X)', (\alpha, \alpha))] \quad \text{and} \quad \theta_2([\alpha]) = [((X, X)'', (\alpha, \alpha))].$$

We can denote $\theta_1([\alpha])$ and $\theta_2([\alpha])$ by $[(\alpha, \alpha)]'$ and $[(\alpha, \alpha)]''$, respectively. Set

$$c = \theta_1 - \theta_2: K_1(R) \rightarrow K_1(K) \quad \text{and} \quad C = \operatorname{Im} c.$$

Then $K_1(R)/ \operatorname{Ker} c \cong C$, where

$$\operatorname{Ker} c = \{[\alpha] \in K_1(R) \mid \theta_1([\alpha]) = \theta_2([\alpha])\}.$$

Since

$$\theta_1([\alpha]) - \theta_2([\alpha]) = [(\alpha\alpha^{-1}, \alpha\alpha^{-1})] = [(1, (1))],$$

we have $C \subseteq \operatorname{Ker} \pi$.

Similar to the case of the K_0 group, we now assume that $s \in J(R)$ and the ring R is complete in the sR-adic topology. Then the relations

$$K_1(K) = \operatorname{Im} \theta_1 \oplus \operatorname{Im} \theta_2 \cong K_1(R) \oplus K_1(R)$$

hold; see Theorem 5.1.1 and Corollary 5.1.3. Then we obtain

$$K_1(K) = \operatorname{Im} \theta_i \oplus C, \ i = 1, 2, \quad \text{and} \quad C = \operatorname{Ker} \pi.$$

Therefore, $\operatorname{Ker} \pi \neq 0$, and this answers the question raised after Theorem 5.4.2. It follows from Corollary 5.4.6 that $B = 0 = A$, where the subgroups B and A are defined before Theorem 5.4.2.

Finally, we point out some possible directions for future research in formal matrix rings (these rings are also called *rings of generalized matrices*). Several such directions are given at the end of Sect. 4.7.

1. In this book, basic information about the groups $K_i(K)$, where $K = \begin{pmatrix} R & M \\ N & S \end{pmatrix}$,
is provided by the groups $K_i(R)$, $K_i(S)$ and their homomorphic images $\mathrm{Im}\,\alpha_i$, $\mathrm{Im}\,\theta_i$,
$i = 1, 2$. It is interesting to find subgroups of $K_i(K)$ which are distinct from $\mathrm{Im}\,\alpha_i$
and $\mathrm{Im}\,\theta_i$. In particular, it is highly desirable to find the structure of the subgroups
$\mathrm{Ker}\,\mu$, $\mathrm{Ker}\,\pi$, $\mathrm{Ker}\,\mu_i$ and $\mathrm{Ker}\,\pi_i$, $i = 1, 2$.

2. The description of various (linear) mappings between formal matrix rings and
the description of mappings from formal matrix rings into various rings is an extensive
substantive area of research. This includes the study of ordinary isomorphisms, Lie
isomorphisms, and derivations of formal matrix rings; e.g., see [13, 18, 67, 70] and
the bibliographies thereof. In [81, Question 4.4], the authors raised the question of
the general form of automorphisms of formal matrices of order 2.

In a series of interesting papers [14, 81, 82, 113, 114], the authors study commuting and centralizing mappings, commuting and centralizing traces, and other
mappings of formal matrix algebras of order 2; the bibliographies of these papers
contain many sources. One of the attractions of these papers are their detailed and
informative introductions. The authors also formulate several concrete problems and
fields of future research. In the review [20], commuting mappings are considered.

It seems to lend perspective to consider listed and other mappings (first of all,
linear mappings) for formal matrix rings of any order n. First of all, it is necessary to
consider formal matrix rings $M(n, R, \Sigma)$ over the ring R. In addition, we can study
the dependence of properties of mappings on multiplier systems Σ.

References

1. Abujabal HAS, Nauman SK (2001) A construction of Morita similar endomorphism rings. J Algebra 235:453–458
2. Abyzov AN (2015) Rings of formal matrices close to regular. Rus Math 59(10):49–52
3. Abyzov AN, Tapkin DT (2015) On certain classes of rings of formal matrices. Rus Math 59(3):1–12
4. Abyzov AN, Tapkin DT (2015) Formal matrix rings and their isomorphisms. Sib Math J 56(6):955–967
5. Abyzov AN, Tuganbaev AA (To appear) Formal matrices and rings close to regular. J Math Sci
6. Amitsur SA (1971) Rings of quotients and Morita contexts. J Algebra 17(2):273–298
7. Anderson FW, Fuller KR (1974) Rings and categories of modules. Springer, New York
8. Asadollahi J, Salarian S (2006) On the vanishing of Ext over formal triangular matrix rings. Forum Math 18(6):951–966
9. Aupetit B, Mouton H, du T, (1996) Trace and determinant in Banach algebras. Stud Math 121(2):115–136
10. Auslander M, Reiten I, Smalø SO (1995) Representation theory of artin algebras. Cambridge University Press, Cambridge
11. Baba Y, Oshiro K (2009) Classical artinian rings and related topics. World Scientific, New Jersey
12. Bass H (1968) Algebraic K-theory. Benjamin Inc, New York, W.A
13. Benkovič D, (2007) Lie derivations on triangular matrices. Linear Multilinear Algebra 55:619–626
14. Benkovič D, Eremita D, (2004) Commuting traces and commutativity preserving maps on triangular algebras. J Algebra 280:797–824
15. Birkenmeier GF, Park JK, Rizvi ST (2013) Extensions of rings and modules. Birkhäuser/Springer, New York, p xx+432
16. Birkenmeier GF, Park JK, Rizvi ST (2002) Generalized triangular matrix rings and the fully invariant extending property. Rocky Mt J Math 32(4):1299–1319
17. Blecher DP, Le Merdy Ch (2004) Operator algebras and their modules. Oxford University Press, Oxford
18. Boboc C, Dăscălescu S, van Wyk L (2012) Isomorphisms between Morita context rings. Linear Multilinear Algebra 60(5):545–563
19. Borooah G, Diesl AJ, Dorsey TJ (2008) Strongly clean matrix rings over commutative local rings. J Pure Appl Algebra 212:281–296
20. Brešar M, (2004) Commuting maps: a survey. Taiwan J Math 8:361–397

© Springer International Publishing AG 2017

P. Krylov and A. Tuganbaev, *Formal Matrices*,

Algebra and Applications 23, DOI 10.1007/978-3-319-53907-2

21. Brown B, McCoy NH (1950) The maximal regular ideal of a ring. Proc Am Math Soc 1(2):165–171
22. Brown WC (1993) Matrices over commutative rings. Marcel Dekker, New York
23. Camillo V, Yu H-P (1994) Exchange rings, units and idempotents. Comm Algebra 22(12):4737–4749
24. Chen H (2001) Stable ranges for Morita contexts. Southeast Asian Math Bull 25:209–216
25. Chen H (2002) Morita contexts with many units. Comm Algebra 30(3):1499–1512
26. Chen H (2003) Strongly π-regular Morita contexts. Bull Korean Math Soc 40(1):91–99
27. Chen H (2005) Ideals in Morita rings and Morita semigroups. Acta Math Sinica 21(4):893–898
28. Climent J-J, Navarro PR, Tortosa L (2011) On the arithmetic of the endomorphism ring End($\mathbb{Z}_p \times \mathbb{Z}_{p^2}$). Appl Algebra Eng Comm Comput 22(2):91–108
29. Cohen M (1982) A Morita context related to finite automorphism groups of rings. Pac J Math 98:37–54
30. Company Cabezos M, Gomez Lozano M, Siles Molina M (2001) Exchange Morita rings. Comm. Algebra 29(2):907–925
31. Davidson KR (1988) Nest algebras. Longman, London
32. Dennis RK, Geller S (1976) K_i of upper triangular matrix rings. Proc Am Math Soc 56:73–78
33. Enochs E, Torrecillas B (2011) Flat covers over formal triangular matrix rings and minimal Quillen factorizations. Forum Math 23:611–624
34. Faith C (1973) Algebra: rings, modules, and categories I. Springer, Berlin
35. Fossum RM, Griffith PA, Reiten I (1975) Trivial extensions of Abelian categories. Lect Notes Math 456:1–12
36. Fuchs L (1970) Infinite Abelian Groups I. Academic Press, New York
37. Gelfand I, Gelfand S, Retakh V, Wilson R (2005) Quasideterminants. Adv Math 193(1):56–141
38. Ghahramani H, Moussavi A (2008) Differential polynomial rings of triangular matrix rings. Bull Iranian Math Soc 34(2):71–96
39. Golan JS (1999) Semirings and their applications. Kluwer Academic Publishers, Dordrecht
40. Goodearl KR (1976) Ring theory. Marcel Dekker, New York
41. Goodearl KR (1979) Von Neumann regular rings. Pitman, London
42. Green EL (1982) On the representation theory of rings in matrix form. Pac J Math 100(1):123–138
43. Gurgun O, Halicioglu S, Harmanci A (2013) Quasipolarity of generalized matrix rings. arXiv:1303.3173v1 [math.RA] 13 Mar 2013
44. Gurgun O, Halicioglu S, Harmanci A (2013) Strong J-cleanness of formal matrix rings. arXiv:1308.4105v1 [math.RA] 19 Aug 2013
45. Han J, Nicholson WK (2001) Extensions of clean rings. Comm. Algebra 29(6):2589–2595
46. Hao Z, Shum K-P (1997) The Grothendieck group of rings of Morita contexts. In: Proceedings of the 1996 Beijing international conference on group theory, pp. 88–97
47. Haghany A (1995) Morita contexts and torsion theories. Math Jpn 42(1):137–142
48. Haghany A (1996) On the torsion theories of Morita equivalent rings. Period Math Hung 32:193–197
49. Haghany A (1999) Hopficity and co-hopficity for Morita contexts. Commun Algebra 27(1):477–492
50. Haghany A (2002) Injectivity conditions over a formal triangular matrix ring. Arch Math 78:268–274
51. Haghany A, Varadarajan K (1999) Study of formal triangular matrix rings. Commun Algebra 27(11):5507–5525
52. Haghany A, Varadarajan K (2000) Study of modules over formal triangular matrix rings. J Pure Appl Algebra 147(1):41–58
53. Haghany A, Varadarajan K (2002) IBN and related properties for rings. Acta Math Hung 94(3):251–261
54. Harada M (1965) On semiprimary PP-rings. Osaka J Math 2:153–161

55. Harte RE, Hernandez C (2005) Adjugates in Banach algebras. Proc Am Math Soc 134(5):1397–1404
56. Henriksen M (1974) Two classes of rings generated by their units. J Algebra 31(1):182–193
57. Herstein IN (1965) A counter-example in Noetherian rings. Proc Nat Acad USA 54:1036–1037
58. Hirano Y (2001) Another triangular matrix ring having Auslander-Gorenstein property. Commun Algebra 29:719–735
59. Horn RA, Johnson ChR (2012) Matrix analysis. Cambridge University Press, Cambridge
60. Huang Q, Tang G (2014) Quasipolar property of trivial Morita context. Int J Pure Appl Math 90(4):423–431
61. Huang Q, Tang G, Zhou Y (2014) Quasipolar property of generalized matrix rings. Commun Algebra 42(9):3883–3894
62. Iwanaga Y, Wakamatsu T (1995) Auslander-Gorenstein property of triangular matrix rings. Commun Algebra 23(10):3601–3614
63. Kashu AI (1983) Radicals and torsions in modules. Stiinta, Chisinau
64. Kashu AI (1987) On localizations in Morita contexts. Math USSR Sbornik 61(1):129–135
65. Kelarev AV (2002) Ring constructions and applications, vol 9. Series in algebraWorld Scientific Publishing Co., Inc, River Edge, NJ
66. Kim KH (1982) Boolean matrix theory and applications. Marcel Dekker, New York
67. Khazal R, Dascalescu S, van Wyk L (2003) Isomorphism of generalized triangular matrix rings and recovery of tiles. Int J Math Math Sci 9:533–538
68. Kovaks I, Silver DS, Williams SG (1999) Determinants of commuting-block matrices. Am Math Mon 106(10):950–952
69. Krylov PA (To appear) Determinants of generalized matrices of order 2. J Math Sci
70. Krylov PA (2008) Isomorphisms of generalized matrix rings. Algebra Logic 47(4):258–262
71. Krylov PA (2013) The group K_0 of a generalized matrix ring. Algebra Logic 52(3):250–261
72. Krylov PA (2014) Calculation of the group K_1 of a generalized matrix ring. Sib Math J 55(4):639–644
73. Krylov PA, Mikhalev AV, Tuganbaev AA (2003) Endomorphism rings of abelian groups. Kluwer Academic Publishers, Dordrecht
74. Krylov PA, Tuganbaev AA (2010) Modules over formal matrix rings. J Math Sci 171(2):248–295
75. Krylov PA, Tuganbaev AA (2015) Formal matrices and their determinants. J Math Sci 211(3):341–380
76. Krylov PA, Tuganbaev AA (To appear) Grothendieck and Whitehead Groups of Formal Matrix Rings. J Math Sci
77. Krylov PA, Tuganbaev AA (2008) Modules over discrete valuation domains, vol 43., Walter de Gruyter, de Gruyter expositions in mathematics, Berlin
78. Krylov PA, Yardykov EYu (2009) On projective and hereditary modules over generalized matrix rings. J Math Sci 163(6):709–719
79. Lam TY (1999) Lectures on rings and modules, New York
80. Lambek I (1966) Lectures on rings and modules. Blaisdell, Waltham
81. Li Y-B, Wei F (2012) Semi-centralizing maps of generalized matrix algebras. Linear Algebra Appl 436:1122–1153
82. Liang X, Wei F, Xiao Z, Fošner A (2014) Centralizing traces and Lie triple isomorphisms on triangular algebras. Linear Multilinear Algebra 8(3):821–847
83. Loustaunau P, Shapiro J (1990) Morita contexts. Lect Notes Math 1448:80–92
84. Loustaunau P, Shapiro J (1990) Homological dimensions in a Morita context with applications to subidealizers and fixed rings. Proc Am Math Soc 110(3):601–610
85. Loustaunau P, Shapiro J (1991) Localization in Morita context with applications to fixed rings. J Algebra 143:373–387
86. Marubayashi H, Zhang Y, Yang P (1998) On the rings of Morita context which are some well-known orders. Commun Algebra 26(5):1429–1444
87. McDonald BR (1984) Linear algebra over commutative rings. Marcel Dekker, New York

88. Ming K (2003) On FI-extending modules. J Chungcheong Math Soc 16(2):79–88
89. Morita K (1958) Duality for modules and its applications to the theory of rings with minimum condition. Sci Rep Tokyo Kyoiku Daigaku 6:83–142
90. Müller BJ (1974) The quotient category of a Morita context. J Algebra 28:389–407
91. Müller M (1987) Rings of quotients of generalized matrix rings. Commun. Algebra 15:1991–2015
92. Nauman SK (2004) Morita similar matrix rings and their Grothendieck groups. Aligarh Bull Math 23(1–2):49–60
93. Nicholson WK (1977) Lifting idempotents and exchange rings. Trans Am Math Soc 229:269–278
94. Palmer I (1975) The global homological dimension of semi-trivial extensions of rings. Math Scand 37:223–256
95. Palmer I, Roos JE (1974) Explicit formulae for the global homological dimension of trivial extensions of rings. J Algebra 27:380–413
96. Poole DG, Stewart PN (1995) Classical quotient rings of generalized matrix rings. Int J Math Math Sci 18(2):311–316
97. Rosenberg J (1994) Algebraic K-theory and its applications. Springer, Berlin
98. Rowen LH (1988) Ring theory. Academic Press, San Diego
99. Sakano K (1984) Maximal quotient rings of generalized matrix rings. Commun Algebra 12(16):2055–2065
100. Sands AD (1973) Radicals and Morita contexts. J. Algebra 24:335–345
101. Sheiham D (2006) Universal localization of triangular matrix rings. Proc Am Math Soc 134(2):3465–3474
102. Small IN (1965) An example in Noetherian rings. Proc Nat Acad USA 54:1035–1036
103. Srinivas V (1991) Algebraic K-theory. Birkhauser, Boston
104. Tang G, Li Ch, Zhou Y (2014) Study of Morita contexts. Commun. Algebra 42:1668–1681
105. Tang G, Zhou Y (2012) Strong cleanness of generalized matrix rings over a local ring. Linear Algebra Appl 437:2546–2559
106. Tang G, Zhou Y (2013) A class of formal matrix rings. Linear Algebra Appl 438(12):4672–4688
107. Tapkin DT (2015) Generalized matrix rings and generalization of incidence algebras. Chebyshevskii Sb 16(3):422–449
108. Tuganbaev A (1998) Semidistributive modules and rings. Kluwer Academic Publishers, Dordrecht
109. Tuganbaev A (2002) Rings close to regular. Kluwer Academic Publishers, Dordrecht
110. Tuganbaev A (2009) Ring theory. Arithmetical modules and rings [in Russian] MCCME. Moscow
111. Veldsman S (2007) Radicals of Morita rings revisited. Bul Acad Stiinte Moldova Math 2:55–68
112. Wisbauer R (1991) Foundations of module and ring theory. Gordon and Breach, Philadelphia
113. Xiao Z, Wei F (2010) Commuting mappings of generalized matrix algebras. Linear Algebra Appl 433:2178–2197
114. Xiao Z, Wei F (2014) Commuting traces and Lie isomorphisms on generalized matrix algebras. Operators and Matrices 8:821–847
115. Yang X, Zhou Y (2008) Strong cleanness of the 2×2 matrix ring over a general local ring. J Algebra 320:2280–2290
116. Yu Yardykov E (2008) Simple modules over generalized matrix rings. J Math Sci 154(3):446–447
117. Zhou Zh (1993) Semisimple quotient rings and Morita context. Commun Algebra 21(7):2205–2210

Index

© Springer International Publishing AG 2017
P. Krylov and A. Tuganbaev, *Formal Matrices*,
Algebra and Applications 23, DOI 10.1007/978-3-319-53907-2

Printed in the United States
By Bookmasters